The Guide to Oilwell Fishing Operations

Tools, Techniques, and Rules of Thumb

The Guide to Oilwell Fishing Operations
Tools, Techniques, and Rules of Thumb

Second edition

by

Joe DeGeare

AMSTERDAM • BOSTON • HEIDELBERG • LONDON
NEW YORK • OXFORD • PARIS • SAN DIEGO
SAN FRANCISCO • SINGAPORE • SYDNEY • TOKYO
Gulf Professional Publishing is an imprint of Elsevier

ELSEVIER

Gulf Professional Publishing is an imprint of Elsevier
225 Wyman Street, Waltham, MA 02451, USA
The Boulevard, Langford Lane, Kidlington, Oxford, OX5 1GB, UK

Second edition **2015**

Notices

Knowledge and best practice in this field are constantly changing. As new research and experience broaden our understanding, changes in research methods, professional practices, or medical treatment may become necessary.

Practitioners and researchers must always rely on their own experience and knowledge in evaluating and using any information, methods, compounds, or experiments described herein. In using such information or methods they should be mindful of their own safety and the safety of others, including parties for whom they have a professional responsibility.

To the fullest extent of the law, neither the Publisher nor the authors, contributors, or editors, assume any liability for any injury and/or damage to persons or property as a matter of products liability, negligence or otherwise, or from any use or operation of any methods, products, instructions, or ideas contained in the material herein.

Library of Congress Cataloging-in-Publication Data
DeGeare, Joe, author.
  The guide to oilwell fishing operations : tools, techniques, and rules of thumb / by Joe P. DeGeare. – 2nd edition.
    pages cm
  Includes bibliographical references and index.
  ISBN 978-0-12-420004-3 (hardback)
1. Oil wells–Equipment and supplies–Fishing. I. Title.
  TN871.2.D37 2015
  622'.3382–dc23

                                                          2014034819

British Library Cataloguing in Publication Data
A catalogue record for this book is available from the British Library

For information on all **Gulf Professional Publishing** publications
visit our website at http://store.elsevier.com

This book has been manufactured using Print On Demand technology. Each copy is produced to order and is limited to black ink. The online version of this book will show color figures where appropriate.

ISBN: 978-0-12-420004-3

I dedicate this book to my Lord and Savior, Jesus Christ.

Let all that I am praise the Lord; may I never forget the good things he does for me.
Psalms 103:2

There is one specific individual that I also dedicate this book to, and that is Gerald Lynde. Gerald went to be with our Lord in 2013, and he will truly missed by many.

I had the pleasure of working with Gerald for many years, and he taught me countless things over those years, but a few them stick out. Never be ashamed of your walk with God; actions speak louder than words; always look for the good in people; and there is always a way to recover a fish. Gerald held dozens of patents, and he was the key inventor of the Metal Muncher patent, which is still the benchmark today for milling.

I would also like to dedicate and give thanks to all of the past and future fishing tool operators, for their innovative thoughts and long hours of work that have made the fishing tool business an exciting and challenging part of the oil industry.

Contents

Disclaimer

All of the contents in this book have been carefully considered and prepared as a matter of general information. The contents are believed to represent situations and conditions reliably that have or could occur, but they are not represented or guaranteed as to their accuracy or application to other conditions or situations. Further, the author has no knowledge or control of their interpretation. Therefore, the content and interpretation and recommendations made in connection within this book are represented solely as a guide and rules for readers' consideration, investigation, and verification. No warranties of any kind, whether expressed or implied, are made in connection therewith. Readers are specifically cautioned, reminded, and advised that any use or interpretation of the contents and resulting use or application are made at their sole risk. In considering the contents, any reader agrees to indemnify and hold the author harmless from all claims and action for loss, damage, death, or injury to persons or property.

Disclaimer

Preface

Since the first edition of *The Guide to Oilwell Fishing Operations* was released in 2003, a few things have changed within the oil industry. There has been an increase in horizontal drilling and a few technology updates have occurred; but as always, fishing is still being done. (We have strived to update all the relevant technology for the second edition of the book; however, we recommend that you always talk to a local fishing tool provider to check on the latest technology updates.) The use of coiled tubing has occurred much more frequently with the increase in horizontal drilling and fracking in the North America market. However, the need for fishing tool supervisors and fishing equipment has not changed. Sooner or later, whether drilling a new well or doing a workover on an existing wellbore, you are likely to have a need for fishing, thanks to fatigue of equipment, hole conditions, planned jobs in workover operations, and human error.

Sound guidelines for successful fishing jobs have evolved over decades of accumulated experience and technological advances. These guidelines should be carefully considered when you have a fishing job to do.

The goal of this book is to offer an overview of these basic guidelines, current fishing practices, and rules of thumb. The text provides a review of the most frequently encountered problems that call for fishing in cased holes, open holes, casing exits, and subsea plugs and abandonment. Tools and techniques for overcoming these problems are described, including a chapter on thru-tubing fishing methods.

Although David Haughton and Mark McGurk, two of the original coauthors, were not able to play a role in the second edition of this book, it would not have been possible without them, and I am truly grateful for there roles in making this happen.

Brian Bernier and Michael R. Reilly have played a large part in this second edition. Their years of down-hole fishing experience allowed them to make key contributions and give an immense amount of support. Brian brings more than 40 years of oilfield and fishing experience in Canada, Latin America, Europe, and the Far East. Michael also has more than 40 years of experience in the fishing industry, and he has worked throughout Europe and the Middle East. Both of them can be reached through LinkedIn.com.

Also, if not for the major service and downhole tool providers that exist today, this book would not be possible. Their continuous development of new technology has brought the fishing tool industry to new heights. I would like to give Baker Hughes, National Oilwell Varco, Logan Oil Tools, and EV Downhole Video an immeasurable thank you. Each of the companies also played a key role in supplying technical information, as well as the captions for many of the figures in this book.

Joe P. DeGeare
San Antonio, Texas

CONVENTIONAL FISHING

In oil-field operations, *fishing* is the technique of removing lost or stuck objects from the wellbore. The term *fishing* is taken from the early days of cable-tool drilling. At that time, when a wireline would break, a crew member put a hook on a line and attempted to catch the wireline to retrieve, or "fish for," the tool. Necessity and ingenuity led these oil-field fishers to develop new "bait." The trial-and-error methods of industry's early days built the foundation for many of the catch tools used currently. Fishing jobs fall into three categories: open hole, when there is no casing in the area of the fish; cased hole, when the fish is inside casing; or thru-tubing, when it is necessary to fish through the restriction of a smaller pipe size (tubing).

Fishing-tool companies have kept pace with the oil industry's rapid development and the deployment of new technology. Today, many are capable of fishing successfully in well depths exceeding 20,000 ft., in high-angle and horizontal wellbores, and in deep water.

A fish can be any number of things, including stuck pipe, broken pipe, drill collars, bits, bit cones, dropped hand tools, sanded-up or mud-stuck pipe, stuck packers, or other junk in the hole. Washovers, overshot runs, spear runs, wireline fishing, stripping jobs, and jar runs are among the many fishing techniques developed to deal with the different varieties of fish.

Because there are so many kinds of fish and fishing jobs, many different tools and methods can be used. Some of them are very simple; others are extremely complex. No two fishing jobs are alike, yet many are similar. A seasoned fishing-tool supervisor will draw from the experience gained from many jobs, as well as the expertise of fellow fishing tool operators.

Fishing jobs are very much a part of the planning process in drilling and workover operations. With the increasing cost of rig time and deeper, more complicated wells, operators will often budget for fishing operations. When a fishing operation is planned for a workover, the operator will work closely with a fishing-tool company to design a procedure and develop a cost estimate. Taking into account the probability of success, the cost of a fishing job has to be less than the cost of redrilling or sidetracking the well for it to make economic sense.

Fishing can be thought of as a risk management strategy. When used successfully, it can save a well. Because fishing is more of an applied skill supported by experience than an exact science, there will be more than one possible solution for a given problem. A clear understanding of the problem, the equipment used to solve the problem, and the best-fitting solution will lead to a successful operation.

This experience usually points to a specific approach when all factors are considered. Although no fishing job technique can be guaranteed, the combination of experienced personnel and continuing advances in fishing-tool technology usually offers an option with a good probability of success.

To maximize this probability, properly planning a fishing job is most important. Preplanning meetings should be held that include everyone involved in the job, such as fishing-tool operators or supervisors, mud-company personnel, rig personnel, electric-wireline company representatives (where applicable), and any others who might become involved. It is much cheaper to determine that a certain procedure will not work *before* doing it.

THRU-TUBING FISHING

The increased use of coiled tubing in the last 30 years has led to many technological advancements in thru-tubing workover applications. These include cleanouts, acid stimulation, milling, underreaming, cutting, and coiled-tubing conveyed thru-tubing fishing systems. The ability to perform these operations without having to pull the production string has provided the operator with a cost-effective alternative to conventional rig workovers. Coiled tubing conveyance also allows remedial operations to be completed without having to kill the well, which eliminates possible formation damage from heavyweight kill fluids in the well. Operations using coiled tubing are usually completed in a much shorter time than conventional rig workovers, which means that the shut-in time of the well is reduced, resulting in less loss of production.

With the ever-increasing horizontal drilling throughout the oil industry, coiled tubing operations have become more acceptable for milling of composite bridge plugs after a fracking operation, as well as removal of ball drop seats from a sleeve system. Extended reach wells can require the need for assistance in getting coiled tubing to the desired depth, which can be done with the use of an agitator tool, as well as friction reducers and bead systems. You should always recommend that the coiled tubing service company run a coiled tubing modeling software program. This will give you a better grasp of where the coiled tubing will friction out and you will not be able to go any further.

Early thru-tubing fishing systems were composed of tools designed for wireline conveyance and did not take advantage of the attributes of the coiled tubing. These tools did not allow circulation through them, and early tools that had been modified to allow circulation restricted flow paths. Also, these tools did not have the tensile strength required to handle the impact loads of the jarring systems being developed for coiled tubing use. In addition, some tools developed for other coiled tubing services, such as inflatable-packer operations, proved inadequate for fishing applications. Some tools had to be designed specifically for thru-tubing fishing operations. A dramatic evolution in thru-tubing fishing-tool technology and design has occurred in the last decade, and individual tool components can now be assembled to suit much more demanding and varied applications.

Thru-tubing fishing systems that run on coiled tubing are used to retrieve many different types of fish. These include coiled tubing conveyed bottom-hole assemblies (BHAs) that have been disconnected, stuck flow-control devices in landing nipples that cannot be retrieved with wireline, inflatable bridge plugs, wireline lost in the hole, and coiled tubing itself. If wireline fishing is unsuccessful, coiled tubing conveyed fishing gives the operator another alternative before a conventional rig workover is required.

Chapters 22 to 30 will outline the specific tools and techniques used in coiled-tubing conveyed thru-tubing fishing applications and services. Tool-string hookup design will also be discussed as applied to common thru-tubing fishing applications being carried out today.

This book will not make you a fishing expert, but it will give you a basic understanding of fishing, fishing tools, and fishing problems that you may encounter. With this knowledge, you should be better prepared to make logical decisions when fishing becomes necessary.

USING COMMUNICATION TO AVOID HAZARDS

Like many other oil-field operations, fishing jobs bring together rig, operating-company, and service-company personnel who may not work closely together every day. When such a group is formed to solve the complex problems that fishing jobs can present, the importance of clear and precise communication cannot be overemphasized. Never assume that people understand the explanation or description of a problem. Because fishing jobs can be hazardous, it is critical to make sure that all descriptions of the problem and plans for its solution are thoroughly understood by all the parties involved.

To avoid hazards, the following steps should be followed prior to and during a fishing job. Remember that these actions can only be performed successfully by employing clear and accurate communication among all parties:

- Collect complete and accurate information.
- Notify all parties involved, such as the fishing-tool company, mud company, and wireline company. It is imperative that all parties cooperate and communicate at all times. This is the most important factor in a successful fishing operation, and it is only through effective communication that the individuals involved will be able to select the proper tools and methods to do the job in the safest, most cost-effective manner.
- Every effort should be made to recover something or to otherwise improve the situation on each trip into the hole. Misruns waste money, and additional mishaps are possible with every additional trip into the hole.
- Drawings (including dimensions) should always be made of everything run into the wellbore. This responsibility should not be left to service-company personnel alone. Operating-company personnel should also make independent measurements and sketches. Keeping track of the accurate dimensions of all equipment is critical for economical fishing.
- If a large or unusual tool or downhole assembly is being run, a contingency plan should be created to fish it. Always ask these questions: Can this be fished? Can it be washed over? Do I have the tools available to fish it? What is the risk of the fishing tools becoming stuck or lost based on the conditions of the hole?
- Confirm that all tools will work properly downhole prior to running them, either by surface testing or by having the service company supply copies of test and inspection reports.
- When running any fishing tool into a well, use a moderate speed. Most fishing tools are designed to go over and around the fish. For a tool to do this, it has to be larger than the fish diameter. In most cases, this makes the tool close to the size of the hole. If the tool is run at a fast speed while going into the hole, it will act as a piston and cause excess pressure below it, which can cause lost circulation. If a space in the well is hit, the tool might wedge so tightly into it that it cannot be pulled out.

- Caution should be taken when pulling fishing equipment out of the hole. Always trip out slowly so the well is not swabbed, which could possibly create a blowout. Follow this procedure with both cased and open holes.
- When fishing retrievable packers, keep in mind that the sealing element will not return to normal size for several hours. This close tolerance can cause problems, such as swabbing the wellbore or hanging up in casing couplings if there is debris on top of the packer.
- Always look at the bottom of the pipe when it is removed from the wellbore. A good example is inspecting the parted joint when pipe has been jet-cut. The flare on the recovered piece of pipe can be measured, which will guide the decision to mill with a milling tool or with a hollow mill-container run with an overshot. This practice should include not only fishing trips, but also when pipe has parted for other reasons. In cases of twist-offs or other failures, the dimensions and configuration of the bottom of the parted joint will provide useful information for fishing.
- Fishing tools are designed to do a particular job, but no single tool is a cure-all. These tools should not be treated roughly to engage a fish. If this is necessary, then something else is wrong. Getting overly aggressive with any tool will only compound your problem.

You should always be prepared and discuss contingency plans with everyone involved. Having none-productive time (NPT) is very costly for everyone on a well site and can make or break the overall job. A contingency plan could consist of an agreement among all parties on what to do next, having extra tools on location at all times, and schematics of equipment that is ready if it is needed to be run into the hole.

Good communication, common sense, and experience will maximize the probability of fishing success and minimize its hazards.

THE ECONOMICS OF FISHING

The most economical fishing job is the one not performed. However, even though drilling or workover plans can be carefully formulated to anticipate problems that could result in fishing, unpredictable factors can and do come into play. Human error, unknown hole conditions, metal fatigue in tubulars, junk in the hole, and faulty equipment are only a few of these.

Fishing is the term for procedures used to retrieve or remove from the wellbore stuck pipe, drill collars, parted tubulars, stuck packers, parted or stuck wireline, and other lost or failed equipment. When these conditions develop, drilling, workover, and completion operations cease, and fishing must be completed before normal operations can resume. The scope and duration of the problem and the efficiency of the solution both have an economic impact on the project.

Fishing should be an economical solution to a problem in the well. A shallow hole with little rig time and equipment investment can justify only the cheapest fishing operation. Before starting an extensive fishing job, you need to consult with all parties involved, such as geologists, reservoir engineers, and others responsible for a well. This may determine whether fishing is warranted and provide guidance concerning the appropriate course of action. For example, the geological information found in the well may indicate that reserves above the fish may be sufficient to justify completion without fishing. You may also find after discussion that doing an open-hole side track or a cased-hole casing exit would be more cost-effective then tackling a complicated fishing job.

There are several papers, studies, formulas, and models that help in making the economic decision to fish or not to fish, and if so, for how long. All have merit, and most major operating companies have their own formulas with which to choose among them. However, so many factors affect the decision that creating a standard checklist applicable to all situations would be impossible. Fortunately, advances in the technology and methods of fishing, milling, and sidetracking, along with a large database of information on fishing operations, have made making these decisions easier for operating companies.

Probability factors are useful in determining the time to be spent fishing. No two fishing jobs are exactly alike, but probability percentages can be derived from similar situations. Decision trees with associated costs should be established for drilling and workover programs in which there are multiple wells and similar situations.

Experience, good judgment, a careful analysis of the problem, and effective communication among all parties will lead to a return to normal drilling, completion, or workover operations with the least amount of lost time and money.

THE CARDINAL RULES OF FISHING

Various general rules and procedures apply to most fishing situations. Operations, downhole equipment, formations, and human error can cause fishing to be necessary, and they also can restrict or prevent completion of a fishing job. Following the rules discussed in this chapter will help ensure the greatest chance of a successful fishing job.

All related factors are important, but some warrant additional emphasis. Competent, experienced personnel are of prime importance in these enterprises, as they can perform most fishing operations, including selecting and supervising a fishing specialist (if required), or even prevent fishing altogether. Always be prepared, since fishing often occurs when least expected. Know what to do, learn all the details about the situation, and take immediate action. Expedite operations and know when to quit or take alternate actions, such as sidetracking, redrilling, or abandonment. Perform job analysis after the fact to prevent a future occurrence or to learn how to handle it more efficiently. Observe the general guides to fishing and the common drilling or workover operations that frequently cause fishing. The keys to a successful fishing job are described in this chapter.

EVALUATE

Evaluate the situation. What is in the hole, and where is it? What are the chances of fishing it out? Evaluate the well records and field history. Gather ideas from the fishing-tool supervisor, tool pusher, drilling/production supervisor, engineer, and drillers. Examine alternative approaches.

Always use safe and proven practices. There may be several workable options on a given job, but a proven method offers the fewest surprises. Also, think about how each step (successful or not) would affect the next one. In addition, it is critical to keep track of what goes into the hole, how it is used, and the results of each run.

COMMUNICATE

The importance of effective communication is discussed elsewhere in this book (including Chapter 2), but it should be underscored here. Communication is the key to success, and it cannot be taken for granted at any time. The following steps should be taken prior to and during a fishing job, and the information obtained thereto should be shared with all parties involved in the job:

- Collect complete and accurate information about the situation.
- Notify the fishing-tool company personnel with enough time to allow them to research the problem, ship the proper tools, and prepare for alternative approaches.

- Ensure that all parties involved understand the situation and agree upon the procedures to be used.
- As the job progresses, keep all parties fully informed. Provide progress reports on topics including fishing success, problems encountered, analysis of those problems, alternative plans developed, and additional equipment needed.

GATHER INFORMATION

Some key factors to be considered, as well as information to be gathered and recorded, during a fishing job are listed next. It is extremely important to record data completely and accurately. Do not restrict information gathering. If additional data would be useful, it should be acquired. Fishing jobs never fail because personnel knew too much.

- Record outside diameters (ODs), inside diameters (IDs), and length of fishing string, and make drawings (see Figure 4-1). Pay special attention to all IDs, drill pipe tube IDs, connection IDs,

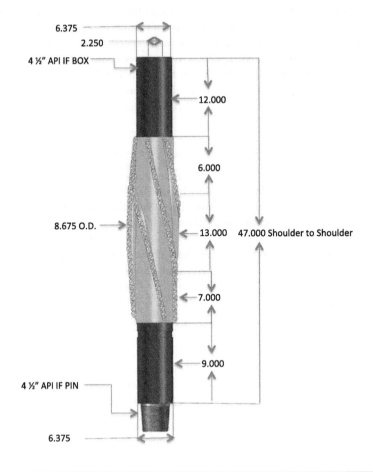

FIGURE 4-1

Example of recording dimensional data for tools being run downhole.

and all tools run in the bottom hole assembly, as they may require ball or wireline tools to be run through them.

- Discuss the job thoroughly with all personnel concerned.
- Know the limitations of the drill pipe and tools on each job.
- Ensure that you have an accurate weight indicator.
- Locate the top of the fish using either wellbore records (e.g., packer), a collar, or a free-point indicator (stuck pipe). Use stretch measurements as a cross-check or by themselves if free-point service is not available.
- Pipe may show to be free in stretch but not free in torque. Torque free-point is recommended for open-hole fishing.
- Combining free-point indicator and freeing pipe by cutting, severing, or backoff may be required to initiate a recovery operation.
- Always leave one or two joints of free pipe above the stuck point when backing off, which will make getting over the top of the fish much easier.
- If the free point is within approximately 100 ft. of the bottom of the casing and open hole, back off up into the casing. It may be impossible to get over the fish if equipment is backed off in an open-hole section below a casing shoe.
- If the stuck point is in a known dogleg or key seat, it may be advisable to pick a straighter section of the hole in which to make a backoff. Also consider the type of formation.
- Determine the depth and condition of the hole and the size of tool joints. These measurements will determine how much back torque it will take to back off a tool joint. A good rule of thumb is that backoff is ¾ round per 1,000 ft. of depth.
- As the size of the drill pipe (fish) decreases, the amount of left-hand torque required to make a backoff increases. For example, to back off a 2⅞-in. OD drill pipe, 1 round per 1,000 ft. may be required.
- If string-shot equipment is not available, consider manual backoff only as a last resort.
- Use a lubricator when running wireline, if possible.

A SIMPLE APPROACH TO THE SOLUTION

Another approach is to use the following method. Most fishing tool supervisors will follow the guide-lines listed next automatically without thinking, and the approach could be incorporated to narrow down the overall guide to a fishing situation. It can be described with the acronyms OD, ID, M, LP, and C:

- OD—Can the fish be caught easily on the outer diameter? This is always the preferential catch to use for the majority of recovery operations.
- ID—If the outer diameter is not available easily, is the inner diameter available without difficulty? This is critical in casing recovery operations.
- M—Will the catch area have to be modified to access the outer or inner diameter? If so, how can this be best accomplished? Keep in mind that the goal is to access the outer or inner diameter.
- LP—Is it necessary to cut the top of the fish, or even the complete fish, into little pieces (by milling or cutting in shorter sections) to access a fishing neck or remove the obstruction completely?
- C—Cement back. There are times when it is necessary to recognize that the reward is not worth the risk of fishing. In these cases, it is best to isolate and leave the upper well bore secure for alternative methods (such as side tracking) or abandon it altogether.

KEEP TRACK OF PIPE TALLIES

Always be aware of the pipe count on a fishing job if it is necessary to lay down pipe. Avoid mixing pipe or drill collars used for fishing with extra pipe on location. A joint count should always be tallied and recorded. Some experienced fishing-tool operators call this process *out-and-in fishing*. Always measure and total all items laid down and measure and total all items picked up. The difference in the totals will equal the amount necessary to pick up or lay down to tag the top of the fish. This serves as a double check if there is difficulty locating the top of a fish.

DO NOT ROTATE THE FISHING STRING

To speed up a trip with a drilling string or workstring, the pipe in the hole is frequently rotated to unscrew the connection. During fishing operations, however, this practice is unacceptable because the fish may be lost. Spinning a fishing tool such as an overshot, spear, magnet, junk basket, or washover pipe frequently causes the fish to be released back into the hole.

DO NOT PULL OUT THE ROPE SOCKET

Most conductor lines are connected to tools or instruments with a rope-socket shear device for a given pull. For some tools, this means of retrieval can be acceptable, but it is dangerous, particularly in open holes when running tools with radioactive sources. The most acceptable method of retrieving these tools is by cutting the line at the surface and stripping over (as described in Chapter 13). If the line is pulled out at the rope socket, the instrument must be retrieved with a catching device that could possibly rupture the canister, allowing the radioactive material to pollute the well fluid.

The fishing of conductor line or swab line with wireline is also an unacceptable practice. Fishing these lines should be done with a pipe workstring. Wireline can ball up and may require considerable pulling to retrieve it.

THE ROLE OF A FISHING TOOL SUPERVISOR

Over time, the role of a fishing tool supervisor has changed. This is very evident in the larger oil company, where numerous staff members, from engineers to team leaders, are involved in a well. There are oil companies that have personnel with backgrounds in most disciplines of wellbore construction, beginning with in-house completions, then progressing to directional drilling, fluids, cementing, and even fishing. All of these disciplines gather to make a team, each which an area of expertise and generally of extensive practical background.

At one time, the fishing tool supervisor and the well-site leader discussed, evaluated, and decided on location, during the operation, the next steps they would pursue. In today's engineered oil wellbore construction and remedial operations, fishing tool supervisors have additional roles and responsibilities. Not only do they have to understand the problem and the tools they use, they must also be able to communicate with others (whether on location or not) about events that occurred, translate and convey indicators from down-hole, and translate and interpret what is found, if anything is recovered. It becomes a difficult task for the fishing supervisor in some cases, but this is where experience helps. It is necessary to be as accurate as possible with these interpretations.

In the current workplace, the fishing tool supervisor must be an effective evaluator, interpreter, communicator, teacher, and student, as well as a tool operator.

PIPE STICKING

Pipe can become stuck during drilling and workover operations, even when preventive measures are taken. When pipe sticking occurs, special tools and expertise are required to avoid expensive, time-consuming, and trial-and-error fishing operations. This chapter presents 10 typical pipe-sticking problems.

SAND STICKING

Sand sticking is a condition usually associated with tubing, although drill pipe and casing can also experience the problem. In tubing, sand sticking is caused by a hole in the tubing, a hole in the casing, or a packer that is not secured. These conditions allow sand to enter the annulus, which prevents the tubing from being pulled. In the case of drill pipe or casing, the well can kick or blow out, causing sand to blow up into the hole and form a sand bridge. If the hydrostatic pressure of the well fluid is not high enough, sand may fall around the pipe.

MUD STICKING

The "setting up" or dehydration of mud in the annulus causes mud sticking (Figure 5-1). In cased holes, temperature can affect some mud additives, causing them to degrade and possibly allowing the barites and solids to settle out. A hole or leak in tubing or casing can allow foreign fluids to contaminate the mud and alter its properties. Typical contaminants are shale, soluble salts, and acid gases. Mud sticking is more often encountered in open holes.

BARITE SETTLING/STICKING/ID RESTRICTING

With the extensive use of Oil Based Mud (OBM), weighted, it is found that American Petroleum Institute (API) barite settles relatively quickly when left static. This settling can impede the internal diameter of drill pipe and certainly will create solid pack bridges in the casing-in-casing wellbore. Recent studies on ultrafine barite have concluded that the settling time needs to be extended. When entering a wellbore, it is critical to know how much fluid is in the annulus and that settling will certainly have occurred if weighted fluid is left behind. This settling will occur in the annular spaces that are created as the inner string traverses the inner diameter of the outer string from side to side due to wellbore curvature and pipe flexibility.

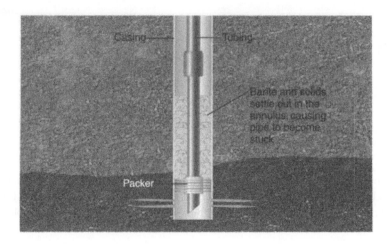

FIGURE 5-1

Mud sticking.

FRICTION STUCK (CASING IN CASING)

As wells are drilled deeper, casing strings are increased in the outer diameter from the surface down. This requires the subsequent casing strings to be maximized in the outer diameter in order to reach production depth, with a production string large enough to flow the well. The relationship between the outer and inner strings is reduced to permit the maximum outer diameter that is permitted for a given inner diameter. The outer casing string traverses the wellbore according to the well geometry and the characteristics of the casing. The subsequent smaller outer diameter casing string has different characteristics, flexibility, and bending tendencies and will traverse a different line.

Once the casing string is in place, the static friction to overcome movement will increase over time as the casing strings settle in contact. The contact points can create excessive static friction, and they prevent long sections of inner casing string from being pulled. This forces the operation into a determination of recoverable length through a series of cuts. Once it is determined that the length of casing can be pulled safely, multiple cuts and recovery operations are needed to pull the casing.

MECHANICAL STICKING

The most frequently occurring mechanical sticking problems are described in the next sections.

STUCK PACKERS OR OTHER DOWNHOLE ASSEMBLIES

Packers and other downhole assemblies have built-in mechanical mechanisms or procedures to allow retrieval. When these mechanisms or procedures fail, the packer or downhole assembly remains in position and cannot be retrieved.

MULTIPLE STRINGS WRAP AROUND

The problem of multiple strings wrapping occurs when tubing strings twist and wind around each other as they are run into the hole (Figure 5-2).

CROOKED PIPE

Crooked pipe is usually the result of dropping a string of tubing, drill pipe, or casing. It may also be caused by weight pressing on a string of stuck pipe (Figure 5-3).

JUNK IN THE HOLE

Another type of mechanical sticking problem includes foreign objects dropped or tools broken off and falling into the hole. The junk can wedge into a collar, tool joint, or bottomhole assembly and stick in the pipe.

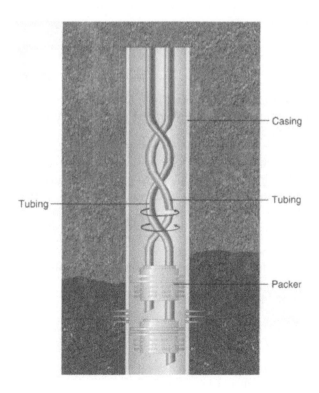

FIGURE 5-2

Mechanical sticking: Multiple strings wrapping around.

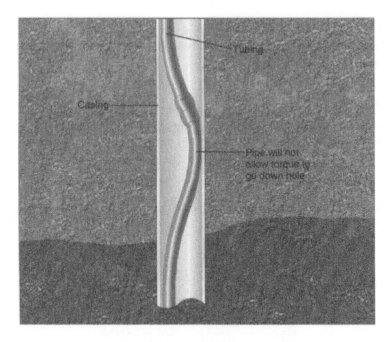

FIGURE 5-3

Mechanical sticking: Crooked pipe.

KEY-SEAT STICKING

A key seat is formed when drill pipe in tension wears a slot (key seat) into the wall of the wellbore during drilling operations (Figure 5-4).

This is more likely to occur when the wellbore deviates from the true vertical position, controlled or otherwise. The worn slot is usually smaller than the wellbore, and the parts of the drill string with the largest diameter—usually drill collars or tool joints—are the most likely to get stuck in a key seat. Even when pipe in a key seat can be moved up and down, it may not be possible to pull a tool joint or drill collar through it.

Four typical indicators of key-seat sticking are as follows:

- Pipe is moving up when it becomes stuck.
- Pipe is stuck at an outside-diameter (OD) enlargement in the drill string.
- Drill string is stuck at the top of the drill collars.
- Circulation of drilling mud is not affected in any way.

CEMENT STICKING

Although cement sticking can result from a mechanical malfunction such as a pump failure or leak in a string of pipe, there are three primary causes (Figure 5-5):

- Displacement has been miscalculated.
- The hole has been washed out as a result of efforts to contain a downhole blowout.
- Efforts have been made to prevent excessive lost circulation.

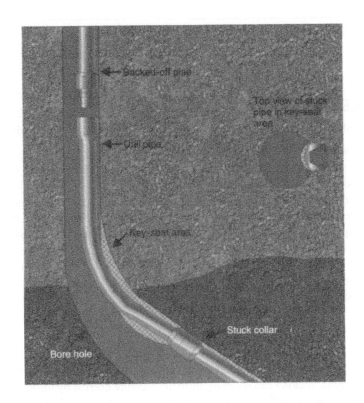

Backed-off pipe

Drill pipe

Top view of stuck pipe in key-seat area

Key-seat area

Stuck collar

Bore hole

FIGURE 5-4

Key-seat sticking.

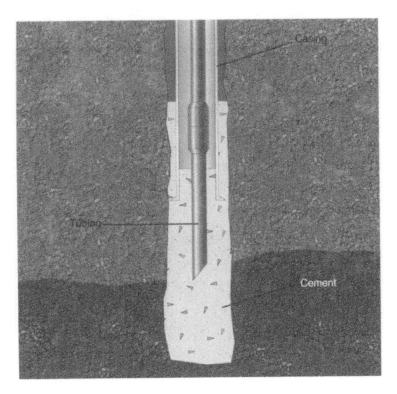

Casing

Tubing

Cement

FIGURE 5-5

Cement sticking.

UNDER-GAUGE HOLE STICKING

The problem of under-gauge hole sticking has several causes, including the following:

- If plastic shale with highly expandable clay content is penetrated when drilling with a fluid of lower hydrostatic pressure than the formation pressure, shale can deform at this differential pressure and close the hole (Figure 5-6).
- Abrasive hole sections will dull drill bits and reduce the bit and stabilizer gauge, making it impossible to get to the bottom of the well with a new bit without reaming. In this situation, tripping in at high speed can jam a full-gauge assembly into an under-gauge hole and get stuck.
- The plastic nature of salt formations may result in stuck pipe. When drilling into salts, stresses will be relieved, and the formation will extrude into the wellbore. Drilling a salt section with oil or mud may cause an under-gauge hole as the weight of the overburden may cause the salt to flow into the wellbore.

DIFFERENTIAL STICKING

Differential sticking occurs in an open hole when any part of the pipe becomes embedded in the mud cake. When this happens, sticking can result because the pressure exerted by the mud column is greater than the pressure of the formation fluids on the embedded section (Figure 5-7). In permeable formations,

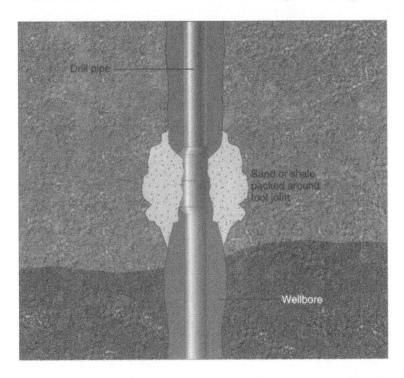

FIGURE 5-6

Under-gauge hole sticking.

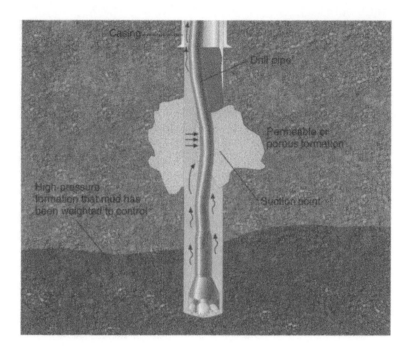

FIGURE 5-7

Differential sticking.

mud filtrate will flow from the well into the rock and build up a filter cake. A pressure differential will be created across the filter cake, which is equal to the difference in the pressure of the mud column and the formation. If the pressure differential is great enough and acts over a sufficiently large area, the pipe can become stuck. Differential sticking can intensify as wall-cake thickness increases (Figure 5-8).

Proper action should be taken immediately if the pipe becomes stuck. Rotation or downward movement of the pipe offers a better chance of breaking the mud seal than does pulling it upward.

In some cases, a differentially stuck string may be freed by reducing the mud weight, which in turn will reduce the differential pressure between the mud column and the permeable zone. This technique should not be used where well control is a problem.

The use of lubricants pumped downhole and placed (spotted) over the affected area can help to free differentially stuck pipe. This is most effective when used in conjunction with jars. In these cases, use overpulling, set-down weights, and torque to free the fish.

Indications of differential sticking typically occur when pipe is not moving, especially while making a connection. The drill string cannot be moved up or down or rotated, but circulation is not affected in any way.

BLOWOUT STICKING

The primary cause of blowout sticking is formation pressure that exceeds the hydrostatic pressure of the well fluid. This can cause a pressure kick or well blowout. This condition is usually the result of insufficient drilling fluid (mud) weight, which has a number of causes, not the least of which is a lack of geological data about the field. Blowout sticking can arise from the following situations (Figure 5-9):

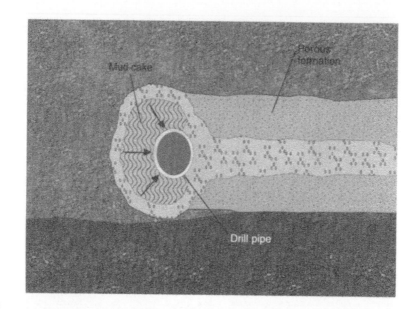

FIGURE 5-8

Increase in wall cake.

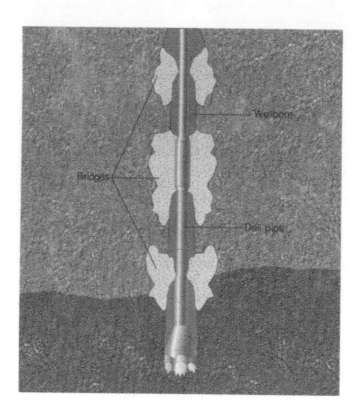

FIGURE 5-9

Blowout sticking.

- **Failure to keep the hole full of fluid.** This can occur if the amount of fluid is not monitored properly while tripping pipe.
- **Swabbing the well.** This can happen when the hydrostatic pressure of the drilling fluid and the formation pressure are close to the balancing point, and the pipe is pulled from the well too quickly. If enough of the hydrostatic pressure of the drilling fluid is relieved, the formation pressure will overcome the drilling-fluid pressure.
- **Mud cut by gas or water.** With this problem, gas or water from the formation can enter the drilling-fluid system and cut the mud weight enough to make the formation pressure exceed the hydrostatic pressure of the drilling fluid.

If a blowout occurs for any of these reasons, sand, shale, other formation debris, and pipe stabilizers (rubber) could be blown up the hole. This debris can bridge over the string of pipe and make it stick.

LOST-CIRCULATION STICKING

Lost-circulation sticking is one of the most frequent problems encountered in drilling operations. It occurs when highly permeable or unconsolidated formations are fractured by the hydrostatic pressure of the drilling fluid (Figure 5-10).

In this situation, drilling fluid can flow freely into shallow, unconsolidated formations because of their high permeability. This fluid flow can cause washouts, sometimes resulting in surface cavities.

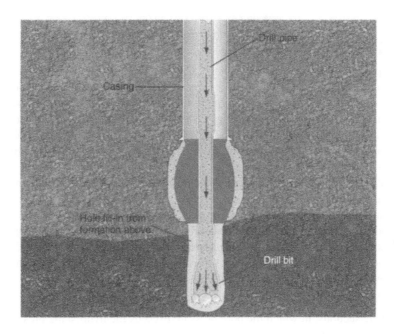

FIGURE 5-10

Lost-circulation sticking.

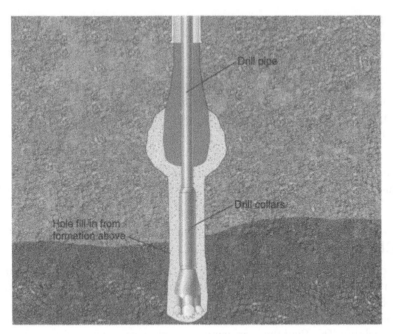

FIGURE 5-11

Sloughing-hole sticking.

SLOUGHING-HOLE STICKING

Shales are the primary culprit in sloughing-hole sticking problems. Categorized as either *brittle* or *sloughing,* the shales fail by breaking into pieces and sloughing into the hole. Then the shales swell as a result of a chemical reaction with water, a process called *hydration.* This reduces the stability of the shale section, causing it to expand perpendicular to the bedding plane and slough off into the wellbore. Massive sloughing forms particles that are too large to fall to the bottom of the hole. Instead, they lodge around the pipe, usually at the drill collars or stabilizers, and cause the pipe string to get stuck. Circulation may also be lost (Figure 5-11).

Shale stability is governed by several factors, including the weight of the overburden, in situ stresses, angle of bedding planes, moisture content, and chemical composition.

Shale sloughing is indicated by large amounts of shale on the shakers at "bottoms up" and also by drag on trips and high levels of fill. Clay platelets, which make up shale, are pushed apart by water, and the formation expands as a result. Any swelling shale is a potential cause of stuck pipe. Some shale will swell rapidly and dramatically. Given sufficient free water, the clay platelets will separate completely, expanding to several times their original volume.

Other causes of sloughing are as follows:

- Pressured shale sections, in which the pore pressure exceeds the hydrostatic pressure.
- Steeply dipping shale beds.
- Turbulent flow in the annulus, which helps promote erosion.

- Ledges breaking off in the wellbore.
- Surge pressures or particles in cavities that slough in when circulation is stopped or the pipe moves.
- Unstable formations extrude into the hole and close around the pipe, while others slough and cause the hole to pack off. Coal is prone to sloughing, salt extrudes, and shale can do either. Uncemented sands and gravel slough into the hole, producing large over-gauge sections and possibly hole-cleaning problems.
- Heavily fractured limestone that results in a succession of boulders falling into the well, jamming around the bottom-hole assembly, and causing the pipe to stick.

TYPES OF FREE POINT TOOLS

The free point tool run on electric line to determine stress and strain in a stuck drill string, casing string, or tubing string has three different configurations. The three different anchoring mechanisms transmit a stress or strain, over-pull, or torque to a strain gauge, which in turn sends various levels of current to the surface for interpretation. These anchors are knives, magnets, or adjustable bow springs. The commonly used configuration is the knife type, as it anchors into the inner diameter with a positive engagement.

The knife type anchors work equally well in terms of tension and torsion. Bow springs are subject to slipping if the tension of the spring is inadequate, and magnets are subject to the condition of the inner wall of the tubular (in particular, casing) that sits in a well for an extended period of time.

DETERMINING THE STUCK POINT

When pipe becomes stuck for any of the reasons described in the previous chapters, the first step is to determine the depth at which the sticking has occurred. Stretch in the pipe should be measured and a calculation made to estimate the distance to the top of the stuck pipe. All pipe is elastic, and all formulas and charts are based on the modulus of elasticity of steel, which is approximately 30,000,000 psi. If the length of stretch in a pipe with a given pull is measured, the amount of free pipe can be calculated or determined from a chart available in data books.

USING STRETCH CALCULATIONS FOR DETERMINING FREE CASING

There are situations where stretch calculations can be reliably used to determine shallow free points on casing. This is more relevant to land operations for abandonment purposes. When determining this method as an indicator of the amount of free casing, the first and foremost consideration is safety. It is paramount to understand the load limits of all the equipment to be used, from the derrick to the spear. The following points give an outline of the operation:

Safe load limits of all components
Buoyed weight of casing
A minimum of three load points
Depth and inclination requirements: less than or equal to 3,000 ft, and less than 25 degrees
Friction factor

Outside diameter (OD)/inside diameter (ID) relationship of casing
Factors for wells in excess of 3,000 ft and 25 degrees
Requirement to follow the correct procedures for pulling and measuring stretch

TUBING, DRILL PIPE, AND CASING STRETCH DATA

The amount of stretch or elongation of oil-well tubular material that results from an applied pulling force is a commonly required measurement. Robert Hooke (1635–1702) discovered that the strain or distortion of a given material is proportional to the stress or force applied, if the elastic limit of the material is not exceeded (known as *Hooke's law*). The elastic limit of a material is the maximum stress that can be developed within it without causing permanent deformation—or *permanent stretch,* in oil-field terms.

The amount of stretch that will occur when a pull force is applied varies with the amount of pull, the length of material being stretched, the elasticity of the material, and its cross-sectional area. These variables are given in the general stretch formula discussed next, provided that the elastic limit of the material is not exceeded.

GENERAL STRETCH FORMULA

$$\Delta L = \frac{F \times L \times 12}{E \times a_s},$$

where ΔL is stretch, in inches; F is pull force, in pounds; L is length, in feet; E is modulus of elasticity, in psi (for steel, $E = 30,000,000$ psi); and a_s is cross-sectional area (wall area or OD area minus ID area for tubular material), in square inches.

Note: It is a common misconception that the rate of stretch for oil-field tubular material is also affected by the grade of steel (j-55, n-80, etc.). This is not true.

Higher grades of steel have greater elastic limits, so they can be stretched farther before reaching their elastic limits than the lower grades can, but the rate of stretch is the same for all grades of steel. The only factors that affect the rate of stretch are those shown in the general stretch formula.

STRETCH TABLES

Stretch tables in this section cover a wide range of sizes and weights of tubing, drill pipe, and casing (Figures 6-1 through 6-5). Columnar tabulations in the tables show OD, weight per foot, ID, pipe-wall cross-sectional area, Stretch Constant (SC), and Free Point Constant (FPC).

DETERMINING STRETCH

The amount of stretch for a particular material is determined using the correct SC, found in the tables via the following formula:

$$\Delta L = F \times L \times SC,$$

OD (in.)	Weight (lb-ft.)	ID (in.)	Wall Area (sq. in.)	SC (in./1.000 lb./ 1.000 ft.)	FPC
1.050 (¾)	1.14	0.824	0.333	1.20120	832.5
	1.20				
1.315 (1)	1.30	1.125	0.364	1.09890	910.0
	1.43	1.097	0.413	0.96852	1032.5
	1.63	1.065	0.467	0.85653	1167.5
	1.70	1.049	0.494	0.80972	1235.0
	1.72				
	1.80				
1.660 (1 ¼)	2.10	1.410	0.603	0.66335	1507.5
	2.30	1.380	0.669	0.59791	1672.5
	2.33				
1.900 (1 ½)	2.40	1.650	0.697	0.57389	1742.5
	2.60	1.610	0.799	0.50063	1997.5
	2.72				
	2.75				
	2.76				
	2.90				
2.000 (2)	3.30	1.670	0.951	0.42061	2377.5
	3.40				
2.063(2 1⁄16)	2.66	1.813	0.761	0.52562	1902.5
	3.25	1.751	0.935	0.42781	2337.5
	3.30				
	3.40				
2.375 (2 ⅜)	3.10	2.125	0.884	0.45249	2210.0
	3.32	2.107	0.943	0.42418	2357.5
	4.00	2.041	1.158	0.34542	2895.0
	4.60	1.995	1.304	0.30675	3260.0
	4.70				
	5.30	1.939	1.477	0.27082	3692.5
	5.80	1.667	1.692	0.23641	4230.0
	5.95				
	6.20	1.853	1.733	0.23081	4332.5
	7.70	1.703	2.152	0.18587	5380.0

FIGURE 6-1

Tubing stretch table.

(courtesy of Baker Oil Tools)

OD (in.)	Weight (lb-ft.)	ID (in.)	Wall Area (sq. in.)	SC (in./1,000 lb./ 1,000 ft.)	FPC
	4.36	2.579	1.268	0.31546	3170.0
	4.64	2.563	1.333	0.30008	3332.5
	6.40	2.441	1.812	0.22075	4530.0
	650				
	7.90	2.323	2.254	0.17746	5635.0
2 875	8.60	2.259	2.484	0.16103	6210.0
(2 ⅞)	870				
	8.90	2.243	2.540	0.15748	6350.0
	9.50	2.195	2.708	0.14771	6770.0
	10.40	2.151	2.858	0.13996	7145.0
	11.00	2.065	3.143	0.12727	7857.5
	11.65	1.995	3.366	0.11884	8415.0
	5.63	3.188	1.639	0.24405	4097.5
	5.75				
	7.70	3.068	2.228	0.17953	5570.0
	9.20	2.992	2.590	0.15444	6475.0
	9.30				
	10.20	2.922	2.915	0.13722	7287.5
3.500	10.30				
(3 ½)	12.80	2.764	3.621	0.11047	9052.5
	12.95	2.750	3.682	0.10864	9205.0
	13.70	2.673	4.010	0.09975	10025.0
	14.70	2.601	4.308	0.09285	10770.0
	15.10	2.602	4.304	0.09294	10760.0
	15.80	2.524	4618	0.08662	11545.0
	17.05	2.440	4.945	0.08089	12362.5
	9.40	3.548	2.680	014925	6700.0
	9.50				
4.000 (4)	10.80				
	10.90	3.476	3.077	0.13000	7692.5
	11.00				
	11.60	3.428	3.337	0.11987	8342.5
	13.40	3.340	3.605	0.10512	9512.5
	12.60	3.958	3.600	0.11111	9000.0
	12.75				
4.500	15.10	3.826	4.407	0.09076	11017.5
(4 ½)	15.50				
	16.90	3.754	4.836	0.08271	12090.0
	19.20	3.640	5.498	0.07275	13745.0

FIGURE 6-2

Tubing stretch table.

(courtesy of Baker Oil Tools)

OD (in.)	Nominal Weight (lb-ft.)	ID (in.)	Wall Area (sq. in.)	SC (in./1.000 lb./ 1.000 ft.)	FPC
2 ⅜	4.85	1.995	1.304	0.30675	3260.0
	6.65	1.815	1.843	0.21704	4607.5
2 ⅞	6.85	2.441	1.812	0.22075	4530.0
	10.40	2.151	2.858	0.13996	7145.0
	9.50	2.992	2.590	0.15444	6475.0
3 ½	13.30	2.764	3.621	0.11047	9052.5
	15.50	2.602	4.304	0.09294	10760.0
4	11.85	3.476	3.077	0.13000	7692.5
	14.00	3.340	3.805	0.10512	9512.5
	13.75	3.958	3.600	0.11111	9000.0
4 ½	16.60	3.826	4.407	0.09076	11017.5
	18.10	3.754	4.836	0.08271	12090.0
	20.00	3.640	5.498	0.07275	13745.0
5	16.25	4.408	4.374	0.09145	10935.0
	19.50	4.276	5.275	0.07583	13187.5
5 ½	21.90	4.778	5.828	0.06863	14570.0
	24.70	4.670	6.630	0.06033	16575.0
6 ⅝	25.20	5.965	6.526	0.06129	16315.0

FIGURE 6-3

Drill pipe stretch table.

(courtesy of Baker Oil Tools)

where ΔL is stretch, in inches; F is pull force, in thousands of pounds; L is length, in thousands of feet; SC is charted SC, in inches of stretch per 1,000 lbs. of pull per 1,000 ft. of length.

EXAMPLE

Determine the amount of stretch for 30,000 lbs. of pull on 6,500 ft. of 2.375 in. OD, 4.7 lb-ft. of 1.995 in. ID tubing as follows:

$$\Delta L = F \times L \times SC$$

$$\Delta L = 30 \times 6.5 \times 0.30675$$

$$\Delta L = 59.8 \text{ in. of stretch}$$

DETERMINING THE FREE POINT

The charted FPC makes it possible to determine very easily the length of pipe being stretched when the amount of pull force and amount of stretch are known. This is commonly referred to as determining

OD (in.)	Nominal Weight (lb-ft.)	ID (in.)	Wall Area (sq. in.)	SC (in./1,000 lb./1,000 ft.)	FPC
4 ½	9.50	4.090	2.766	0.14461	6915.0
	10.50	4.052	3.009	0.13293	7522.5
	11.60	4.000	3.338	0.11983	8345.0
	13.50	3.920	3.836	0.10428	9590.0
	15.10	3.826	4.407	0.09076	11017.5
	16.90	3.740	4.918	0.08133	12295.0
5	11.50	4.560	3.304	0.12107	8260.0
	13.00	4.494	3.773	0.10602	9432.5
	15.00	4.408	4.374	0.09145	10935.0
	18.00	4.276	5.275	0.07583	13187.5
	20.80	4.156	6.069	0.06591	15172.5
5 ½	14.00	5.012	4.029	0.09928	10072.5
	15.50	4.950	4.514	0.08861	11285.0
	17.00	4.892	4.962	0.08061	12405.0
	20.00	4.778	5.828	0.06863	14570.0
	23.00	4.670	6.630	0.06033	16575.0
6 ⅝	20.00	6.049	5.734	0.06976	14335.0
	24.00	5.921	6.937	0.05766	17342.5
	28.00	5.791	8.133	0.04918	20332.5
	32.00	5.675	9.177	0.04359	22942.5
7	17.00	6.538	4.912	0.08143	12280.0
	20.00	6.456	5.749	0.06958	14372.5
	23.00	6.366	6.656	0.06010	16640.0
	26.00	6.276	7.549	0.05299	18872.5
	29.00	6.184	8.449	0.04734	21122.5
	32.00	6.094	9.317	0.04293	23292.5
	35.00	6.004	10.172	0.03932	25430.0
	38.00	5.920	10.959	0.03650	27397.5
7 ⅝	24.00	7.025	6.904	0.05794	17260.0
	26.40	6.969	7.519	0.05320	18797.5
	29.70	6.875	8.541	0.04663	21352.5
	33.70	6.765	9.720	0.04115	24300.0
	39.00	6.625	11.192	0.03574	27980.0

FIGURE 6-4

Casing stretch table.

(courtesy of Baker Oil Tools)

the free point in a string of stuck or anchored pipe. Read the correct FPC from the table for the pipe involved and use it in the following formula:

$$L = \frac{\Delta L \times FPC}{F},$$

where L is the minimum length of free pipe, or length being stretched, in ft.; ΔL is stretch, in inches; F is pull force, in thousands of pounds; FPC is charted FPC.

OD (in.)	Nominal Weight (lb-ft)	ID (In.)	Wall Area (sq. in.)	SC (in./1,000 lb./1,000 ft.)	FPC
(8 ⅝)	24.00	8.097	6.934	0.05769	17335.0
	28.00	8.017	7.947	0.05033	19867.5
	32.00	7.921	9.149	0.04372	22872.5
	36.00	7.825	10.336	0.03870	25840.0
	40.00	7.725	11.557	0.03461	28892.5
	44.00	7.625	12.673	0.03156	31682.5
	49.00	7.511	14.118	0.02833	35295.0
(9 ⅝)	32.30	9.001	9.128	0.04382	22820.0
	36.00	8.921	10.254	0.03901	25635.0
	40.00	8.835	11.454	0.03492	81635.0
	43.50	8.755	12.559	0.03185	31397.5
	47.00	8.681	13.572	0.02947	33930.0
	53.50	8.535	15.546	0.02573	38865.0
(10 ¾)	32.75	10.192	9.178	0.04358	22945.0
	40.50	10.050	11.435	0.03498	20587.5
	45.50	9.950	13.006	0.03076	32515.0
	51.00	9.850	14.561	0.02747	36402.5
	55.50	9.760	15.947	0.02508	39867.5
	60.70	9.660	17.473	0.02289	43682.5
	65.70	9.560	18.982	0.02107	47455.0
(11 ¾)	42.00	11.084	11.944	0.03349	29860.0
	47.00	11.000	13.401	0.02985	33502.5
	54.00	10.880	15.463	0.02587	38657.5
	60.00	10.772	17.300	0.02312	43250.0
(13 ⅜)	48.00	12.715	13.524	0.02958	33810.0
	54.50	12.615	15.514	0.02578	38785.0
	61.00	12.515	17.487	0.02287	43717.5
	68.00	12.415	19.445	0.02057	48612.5
	72.00	12.347	20.768	0.01926	51920.0
(16)	65.00	15.250	18.408	0.02173	46020.0
	75.00	15.124	21.414	0.01868	53535.0
	84.00	15.010	24.112	0.01659	60280.0
(20)	94.00	19.124	26.918	0.01486	67295.0

FIGURE 6-5

Casing stretch table.

(courtesy of Baker Oil Tools)

Note: Because of friction forces, which cannot be determined readily, the actual length of free pipe may be longer than calculated. The formula necessarily assumes the complete absence of friction.

EXAMPLE

Determine the minimum length of free pipe being stretched when a string of 4½-in. OD, 16.60-lb-ft. drill pipe stretches 18.6 in. with an applied pull of 25,000 lbs. as follows:

$$L = \frac{\Delta L \times \text{FPC}}{F}$$

$$L = \frac{18.6 \times 11,017.5}{25}$$

$$L = 8,197, \text{or approximately } 8,200 \text{ ft.}$$

CALCULATION OF SC AND FPC

For any pipe sizes not included in the tabulated stretch chart data, SC and FPC can be calculated as follows:

$$\text{SC} = \frac{0.4}{a_s}$$

$$\text{FPC} = 2500 \times a_s,$$

where a_s is the pipe-wall cross-sectional area, in square inches

EXAMPLE

Determine the SC for 2.375-in. OD, 4.7 lb-ft. tubing that has a pipe-wall cross-sectional area (a_s) of 1.304 square inches as follows:

$$\text{SC} = \frac{0.4}{a_s}$$

$$\text{SC} = \frac{0.4}{1.304}$$

$$\text{SC} = 0.30675$$

EXAMPLE

Determine the FPC for 4½-in OD, 16.60 lb-ft. drill pipe that has a pipe-wall cross-sectional area (a_s) of 4.407 square inches as follows:

$$\text{FPC} = 2,500 \times a_s$$

$$\text{FPC} = 2,500 \times 4.407$$

$$\text{FPC} = 11,017.50$$

STRETCH GRAPHS

This section discusses stretch graphs for 1.660-in. through 7-in. OD external upset or non-upset API tubing in the most common weight and wall thickness for each size (see Figures 6-6 through 6-16).

Connection	OD		Weight Plain End		ID		Wall Thickness		Upset Diamater		Drift Diameter		Washover Size			
													Rec.		Max.	
	in.	mm	lb-ft.	kg/m	in.	mm	in.	mm	in.	mm	in.	mm	in.	mm	in.	mm
TSWP	3 3/8	85.73	10.02	14.89	2.764	70.21	0.305	7.75	N/A	N/A	2.639	67.03	2 1/2	63.5	2 5/8	66.68
TSWP	3 1/2	88.9	12.31	18.3	2.764	70.21	0.368	9.35	N/A	N/A	2.639	67.03	2 1/2	63.5	2 5/8	66.68
TSWP			881	13.09	2.992	76.0	0.254	6.45			2.867	72.82	2.69	68.26	2 7/8	73.03
TSWP	3 5/8	92.08	7.06	10.49	3.240	82.3	0.192	4.88	N/A	N/A	3.115	79.12	3.00	76.2	3 1/8	79.38
TSWP	3 3/4	95.25	9.55	14.19	3.238	82.25	0.256	6.5	N/A	N/A	3.113	79.07	3.00	76.2	3 1/8	79.38
TSWP			10.46	15.55	3.185	80.9	0.283	7.19			3.06	77.72	3.00	76.2	3 1/8	79.38
TSWP	3 13/16	96.84	11.7	17.39	3.187	80.95	0.313	7.95	N/A	N/A	3.062	77.77	3.00	76.2	3 1/8	79.38
TSWP	4	101.6	12.93	19.22	3.340	84.84	1.33	8.38	N/A	N/A	3.215	81.66	3 1/16	77.79	3 1/4	82.55
TSWP	4 3/8	111.13	12.02	17.87	3.826	97.18	0.275	6.99	N/A	N/A	3.701	94.01	3 1/2	88.9	3 3/4	95.25
TSWP			13.8	20.18	3.749	95.22	0.313	1.95			3.624	92.05	3 1/2	88.9	3 5/8	92.08
TSS	4 1/2	114.3	11.3	16.87	4.000	101.6	0.25	6.35	N/A	N/A	3.875	98.43	3 3/4	95.25	3 7/8	98.43
TSWP			13.04	19.38	3.920	99.57	1.29	7.37			3.795	96.39	3 5/8	82.08	3 3/4	95.25
TSWP			14.98	22.26	3.826	97.18	0.337	8.56			3.701	94.01	3 1/2	88.9	3 11/16	93.66
TSS	4 3/4	120.65	17.2	26.04	4.000	101.6	0.375	9.3	N/A	N/A	3.875	98.43	3 3/4	9.25	3 7/8	98.43
TSWP	4 7/8	123.83	11.57	17.2	4.408	112.0	0.233	5.92	N/A	N/A	4.283	108.8	4.00	104.6	4 1/8	104.8
TSWP	5	127.0	14.87	22.1	4.408	112.0	0.296	7.52	N/A	N/A	4.283	108.8	4.00	101.6	4 1/8	104.8
TSWP																
X-LINE			15 00	22.29	4.375+	111.1			5.36	136.1					4 1/4	108.0
X-L1NE			18.00	26.75	4.250·	108.0	0.362	9.19			4.151	105.4			4 1/8	104.8
TSWP			17.93	26.65	4.276	108.6			N/A	N/A						
TSS	5 3/8	136.3	20.2	29.76	4.625	117.5	0.375	9.53	N/A	N/A	4.5	114.3	4 1/4	107.95	4 1/2	114.3
TSWP	5 1/2	139.7	16.87	2.07	4.892	124.3	0.304	7.72	N/A	N/A	4.767	121.1	4 5/8	117.48	4 3/4	120.7
X-LINE			17.00	25.27	4.875	123.8			5.86	148.8						
TSWP			19.81	29.44	4.778	11.4	0.361	9.17	N/A	N/A	4.653	118.2	4 1/2	114.3	4 5/8	117.5
TSWP	5 3/4	146.05	18.18	7.02	5.124	130.2	0.313	7.95	N/A	N/A	4.999	127.0	4 7/8	123.83	5	127.0
TSS			21.53	32	5	127.0	0.375	9.53			4.875	123.8	4 3/4	120.65	4 7/8	123.8
TSS	6	152.4	15.35	22.81	5.5	139.7	0.25	6.35	N/A	N/A	5.375	136.5	5 1/4	133.35	5 3/8	136.5
TSWP			19.64	29.19	5.352	135.9	0.324	8.23			5.227	132.8	5 1/8	130.18	5 1/4	133.4
TSWP			22.81	33.9	5.24	133.1	0.38	9.65			5.115	129.9	5.00	127.0	5 1/8	130.2
TSWP	6 3/8	161.93	24.03	35.72	5.625	142.9	0.375	9.53	N/A	N/A	5.5	139.7	5 3/8	136.53	5 1/2	139.7
TSWP	6 5/8	168.28	23.58	35.05	5.921	150.4	0.352	8.94	N/A	N/A	5.796	147.2	5 5/8	142.88	5 3/4	146.1
X-LINE			24.00	35.67	5.91	150.4			7	177.8						
TSWP	7	177.8	25.66	38.14	6.276	159.4	0.362	9.19	N/A	N/A	6.151	156.2	6.00	152.4	6 1/8	155.6
X-LINE			26.00	38.64	6.276				7.39	187.7						
TSWP	7 1/4	184.15	23.19	34.47	6.624	168.3	0.313	7.95	75	190.5	6.499	165.1	6 3/8	161.96	6 1/2	165.1
TSWP	7 3/8	187.3	28.04	41.68	6.63	168.3	0.38	9.5	N/A	0	6.5	165	6 3/8	161.96	6 1/2	165.1
TSWP																

(Continued)

Connection	OD in.	OD mm	Weight Plain End lb-ft	Weight Plain End kg/m	ID in.	ID mm	Wall Thickness in.	Wall Thickness mm	Upset Diameter in.	Upset Diameter mm	Drift Diameter in.	Drift Diameter mm	Washover Rec. in.	Washover Rec. mm	Washover Max. in.	Washover Max. mm
TSWP	7 5/8	193.7	25.56	37.99	6.97	177	0.33	8.3	N/A	0	6.84	174	6 3/4	171.5	6 7/8	174.6
TSWP			29.04	43.16	6.88	174.6	0.38	9.5			6.75	172	6 5/8	168.3	6 3/4	171.5
TSWP			33.04	49.11	6.77	171.8	0.43	10.9			6.64	169	6 1/2	165.1	6 5/8	168.3
TSWP	8	203.2	29.70	44.14	6.843-	173.8	0.38	9.5			6.75	17.2	6 9/16	166.7	6 11/16	169.9
X-LINE			30.54	45.39	7.25	184.2	0.38	9.5	8.01	204	7.13	181	7	177.8	7 1/8	181
TSWP	8 1/8	206.4	31.04	46.13	7.38	187.3	0.38	9.5	N/A	0	7.25	184	7 1/8	181	7 1/4	184.2
TSWP			35.92	53.39	7.25	184.2	0.44	11.1			4.13	181	7	177.8	7 1/8	181
TSWP	8 3/8	212.7	38.42	57.1	7.19	182.5	0.47	11.9	N/A	0	7.06	179	6 15/16	176.2	7 1/16	179.4
TSWP			33.95	50.46	7.58	192.5	0.4	10.1			7.45	189	7 1/4	184.2	7 3/8	187.3
TSWP	8 5/8	219.1	37.09	55.13	7.5	190.5	0.44	11.1			7.38	187				
TSS			31.1	46.22	7.92	201.2	0.35	8.9			7.8	198			7 11/16	195.3
X-LINE			36	53.51	7.813+	198.5	0.4	10.2	9.12	232	7.7	196			7 5/8	193.7
TSWP	9	228.6	39.29	58.4	7.73	196.2	0.45	11.4			7.6	193	7 9/16	192.1	7 11/16	195.3
TSWP			38.92	57.85	8.150	207	0.43	10.8	N/A	0	7.99	203	7 1/2	190.5	8	203.2
TSWP	9 5/8	244.5	38.94	57.88	8.84	224.4	0.4	10			8.68	220	7 7/8	200	8 3/8	212.7
TSWP			42.7	63.46	8.76	222.4	0.44	11	N/A	0	8.6	218	8 1/2	215.9	8 1/2	215.9
X-LINE			43.5	64.65	8.67	220.1			10.1	257						
TSWP			46.14	68.58	8.68	220.5	0.47	12			8.53	217	8 1/4	209.6		
TSWP	10 3/4	273.1	44.22	65.72	9.95	252.7	0.4	10.2	N/A	0	9.79	249	9 1/2	241.3	9 3/4	247.7
TSWP			49.5	73.57	9.85	250.2	0.45	11.4			9.69	246	9 3/8	238.1	9 5/8	244.5
TSWP	11 3/4	298.5	54.21	80.57	9.76	247.9	0.5	12.6			9.6	244	9 1/4	235	9 1/2	241.3
TSWP			52.57	78.13	10.9	276.4	0.44	11			10.7	272	10 1/8	257.2	10 5/8	269.9
TSWP	12 3/4	298.5	58.81	87.41	10.8	273.6	0.49	12.4			10.6	270	10	254	10 1/2	266.7
TSWP	13 3/8	339.7	49.56	73.66	12	304.8	0.38	9.5	13.5	343	11.8	301	11	279.4	11 1/2	292.1
TSWP			66.11	98.26	12.4	315.3	0.48	12.2	13.75	349	12.3	311	11 1/2	292.1	12	304.8
TSWP	16	406.4	81.97	121.8	15	381.1	0.5	12.6	16.75	426	14.8	377	14 1/4	362	14 3/4	374.6

Note: All strengths at maximum value—apply a safety factor of 2 to the joint tensile yield strength.
● = Recommended makeup torque is 25% of the maximum makeup torque; does not apply to X-line connections.
■ = Ratio of the joint tensile yield strength to the pipe tensile yield strength.
◆ = The internal upset has been reduced.
● = N-80 pipe
◇ = J-55 material.

FIGURE 6-6

Tubing stretch chart for 1.660-in. OD, 2.4-lb-ft. EU or NU API tubing.

(courtesy of Baker Oil Tools)

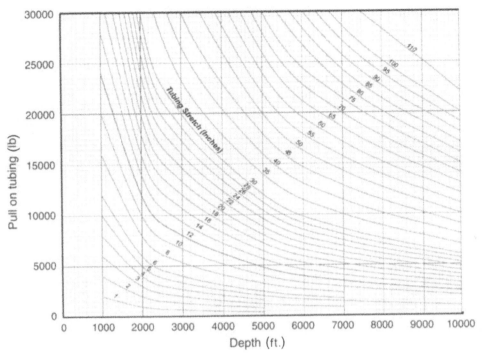

FIGURE 6-7

Tubing stretch chart for 1.900-in. OD, 2.9-lb-ft. EU or NU API tubing.

(courtesy of Baker Oil Tools)

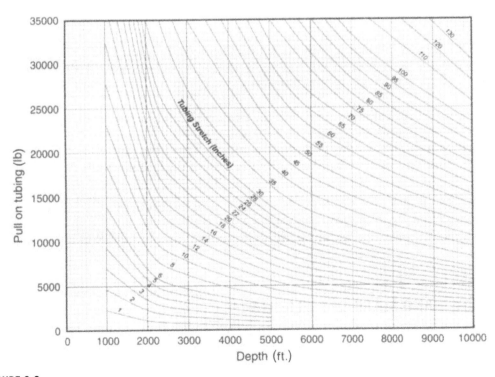

FIGURE 6-8

Tubing stretch chart for 2.062-in. OD, 3.25-lb-ft. EU or NU API tubing.

(courtesy of Baker Oil Tools)

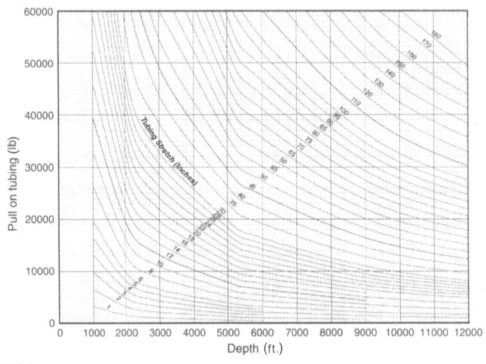

FIGURE 6-9

Tubing stretch chart for 2⅜-in. OD, 4.7-lb-ft. EU or NU API tubing.

(courtesy of Baker Oil Tools)

For tubing having any other cross-sectional wall area, stretch must be determined from the general stretch formula or from the stretch charts included in this section.

Each stretch graph involves only three variables: amount of pull force, depth (or length), and amount of stretch. When any two of the variables are known, the third can be read directly from the graph as follows:

1. If the depth and pull force are known, the amount of stretch can be found.
2. If the depth and stretch are known, the amount of pull can be found.
3. If the pull force and stretch are known, the depth or length of tubing being stretched can be found.

SET-DOWN AND SLACK-OFF WEIGHTS

When a string of tubing is lowered to put weight on the bottom, as in setting a packer, the tubing buckles in the form of a helix, and a significant amount of the applied weight is supported by

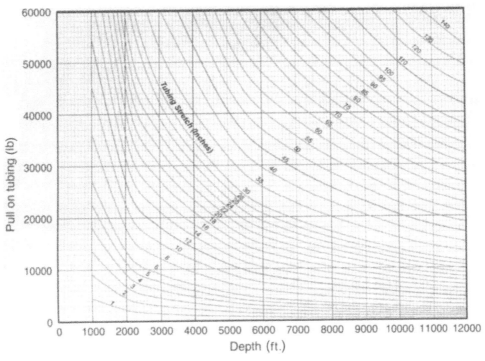

FIGURE 6-10

Tubing stretch chart for 2⅞-in. OD, 6.5-lb-ft. EU or NU API tubing.

(courtesy of Baker Oil Tools)

friction between the tubing and casing. The slack-off graphs found in this chapter (see Figures 6-17 through 6-22) indicate the magnitude of the effect of friction and provide a means of determining the approximate amount of weight applied on the bottom as the tubing is lowered and the weight loss is measured at the surface. Graphs are provided for the most common tubing/casing combinations.

The graphs were developed from mathematical calculations using an assumed average value for the coefficient of friction. They are presented for information only and may not be exactly accurate for any specific case because the coefficient of friction involved may vary from the assumed value; however, actual tests run in a variety of well fluids indicate that the variations are relatively small.

In situations in which the amount of effective tubing weight on the bottom may be marginal or inadequate to completely pack off a set-down-type packer, an attempt should be made to pressure the casing. Pressure in the casing/tubing annulus tends to straighten the tubing and put more weight on the packer. Casing pressure will also increase the pack-off force in the packing element of a partially packed-off set-down-type packer.

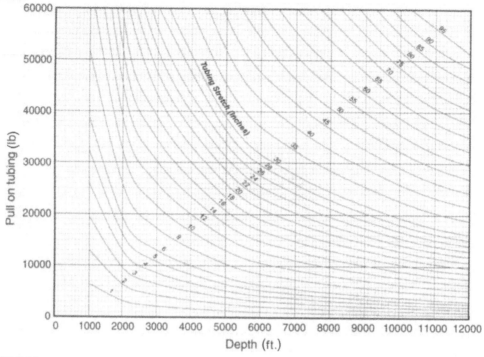

FIGURE 6-11

Tubing stretch chart for 3½-in. OD, 9.3-lb-ft. EU or NU API tubing.

(courtesy of Baker Oil Tools)

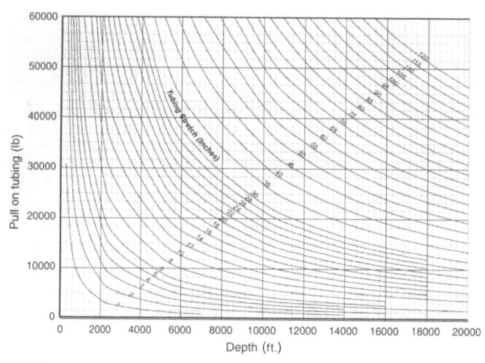

FIGURE 6-12

Tubing stretch chart for 4-in. OD, 10.0-lb-ft. EU or NU API tubing.

(courtesy of Baker Oil Tools)

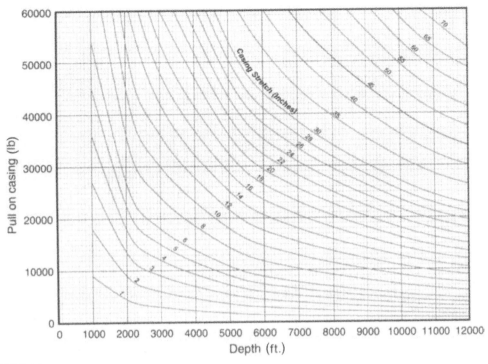

FIGURE 6-13

Casing stretch chart for 4½-in. OD, 12.75-lb-ft. EU or NU API casing.

(courtesy of Baker Oil Tools)

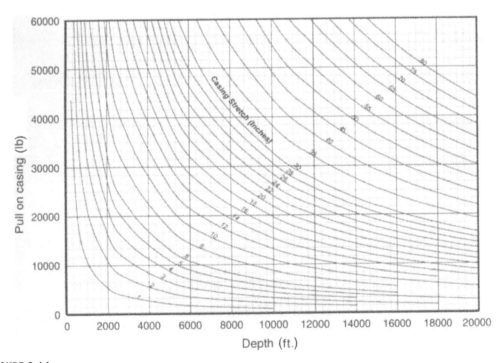

FIGURE 6-14

Casing stretch chart for 5-in. OD, 15.0-lb-ft. EU or NU API casing.

(courtesy of Baker Oil Tools)

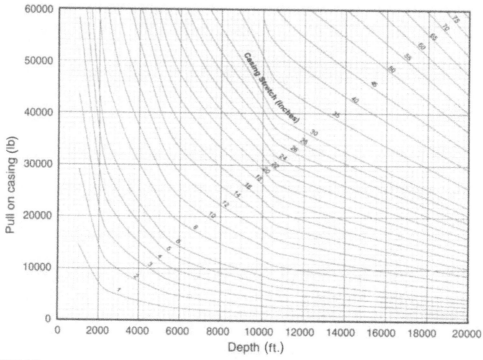

FIGURE 6-15

Casing stretch chart for 5½-in. OD, 20-lb-ft. EU or NU API casing.

(courtesy of Baker Oil Tools)

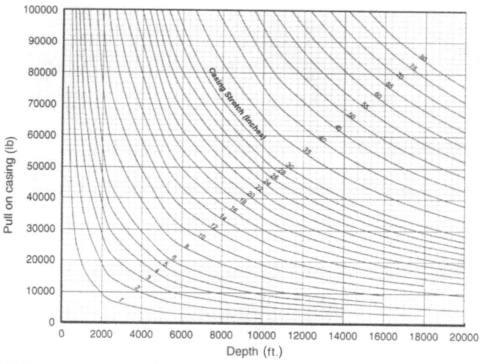

FIGURE 6-16

Casing stretch chart for 7-in. OD, 26.0-lb-ft. EU or NU API casing.

(courtesy of Baker Oil Tools)

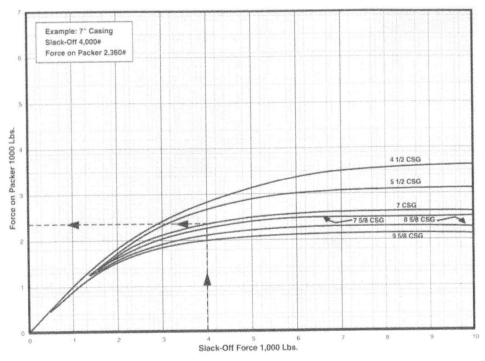

FIGURE 6-17

Weight on packer chart for 1.660-in. OD, 2.4-lb-ft. EU or NU API tubing.

(courtesy of Baker Oil Tools)

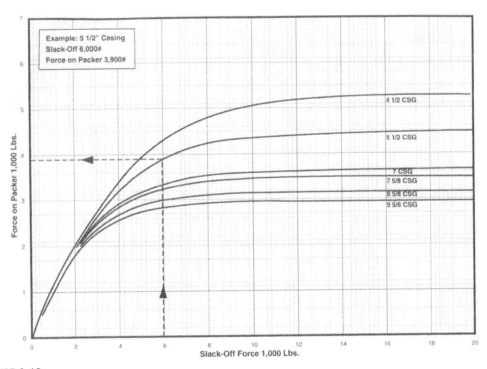

FIGURE 6-18

Weight on packer chart for 1.900-in. OD, 2.9-lb-ft. EU or NU API tubing.

(courtesy of Baker Oil Tools)

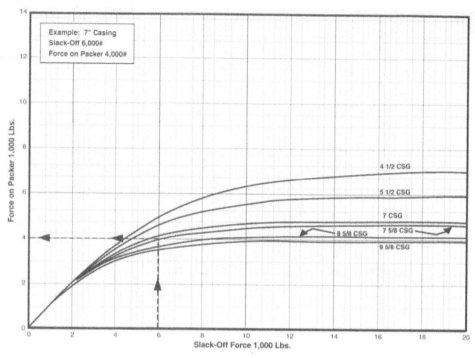

FIGURE 6-19

Weight on packer chart for 2.062-in. OD, 3.25 lb-ft. EU or NU API tubing.

(courtesy of Baker Oil Tools)

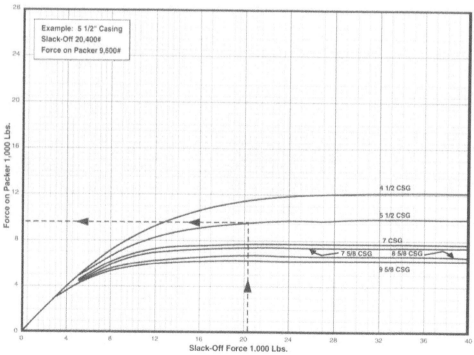

FIGURE 6-20

Weight on packer chart for 2⅜-in. OD, 4.7-lb-ft. EU or NU API tubing.

(courtesy of Baker Oil Tools)

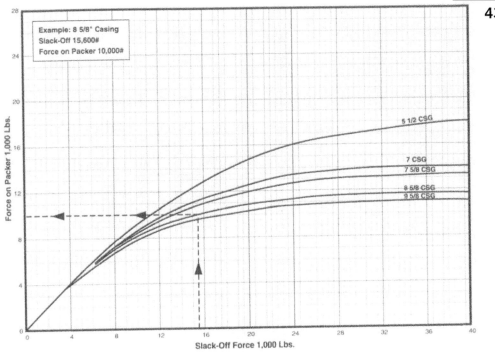

FIGURE 6-21

Weight on packer chart for 2⅞-in. OD, 6.5-lb-ft. EU or NU API tubing.

(courtesy of Baker Oil Tools)

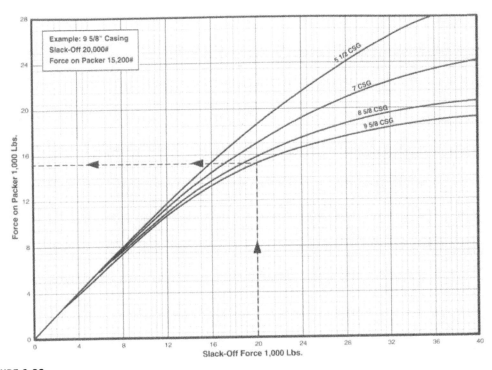

FIGURE 6-22

Weight on packer chart for 3½-in. OD, 9.3-lb-ft. EU or NU API tubing.

(courtesy of Baker Oil Tools)

PARTING THE PIPE STRING

7

After determining the stuck point in a pipe string, the normal procedure is to part the string so that fishing tools, such as jarring strings or washpipe strings, may be run. There are many acceptable methods of parting the pipe string, including the following:

- Mechanical backoff
- String shot assist backoff
- Chemical cutting
- Jet cutting
- Electric-line (e-line) motor-driven cutting
- Mechanical cutting
- High-velocity, abrasive fluid cutting
- Split shots

The method that you choose will depend on the situation that you are in.

The cutting method for each particular job should be selected carefully. For example, only the backoff method leaves threads facing up, so it should be employed if a retrieving tool is to be screwed into the fish. Screwing back into a fish is the preferred method because doing so will restore the pipe to its full strength and ID. If an engaging tool is run, it may be weaker than the pipe or connection, or it may have a larger or smaller ID.

When parting pipe, always leave enough free pipe above the stuck point to act as a guide and provide a catching surface that is long enough for good pulling strength. A sufficient length for these purposes is usually considered to be between a half-joint and two joints. When determining how much free pipe to leave, consider the next operation that you must perform. For example, if you will be washing over inside casing and using a setup where no threads are needed, cutting a half-joint above the stuck point may be adequate. However, if the drill pipe is backed off so that washing over (by using a drill collar spear in the washpipe) can be done, and if there is considerable settling out of solids, extra pipe should be left in the hole.

Some fishing tool operators also like to leave a spare tool joint in place in case the first is damaged by the backoff operation. This will also prevent you needing to fire a string shot in the same joint if the fish cannot be pulled or jarred free. Never leave more pipe above a stuck point than is needed because it contributes to greater washing over if this fishing method is used. If the jarring method is used, extra pipe adds elasticity to the fish.

The methods of conveying cutters into a well have improved so that now there are numerous options, depending on the cutter. In particular, the chemical cutter and jet cutter can be run on coil or on a slick-line, which permits the operator to select an alternative method of conveying the cutter. There are both timed firing heads and pressure-activated firing heads. Thus, with the correct configuration, it is possible to work around the inner-diameter (ID) restrictive cases.

BACKOFF

Backing off is the procedure of applying a left-hand torque to a pipe string while firing a shot of prima cord explosive to unscrew the pipe at a selected threaded joint above the stuck point (Figure 7-1). The resulting explosion produces a concussion that partially unscrews the threads.

The backoff method of parting pipe is probably the most popular, particularly in drill pipe and drill collars, as this is the only method that leaves a threaded connection at the top of the remaining pipe. This makes it possible to screw back into the fish with a workstring when using fishing tools such as jars. Tool joints used on drill pipe, drill collars, and other drilling tools have coarse threads, large tapers, and a metal-to-metal seal with flat surfaces or faces. These characteristics make the backoff method attractive in such situations.

Tubing or coupled pipe does not lend itself to backoff in the same way. Threads are usually fine (at least eight per inch) with only a slight taper (such as ¾ in. per foot), and the threads are in tension with a high degree of thread interference. In spite of these differences, backoff is still a popular method of parting tubing.

In wells that have a mud-type completion fluid that has "settled out" over the years, a good case can be made for backing off the tubing and screwing it back together. Rather than picking up long strings of washover pipe, it may be more economical to back off the tubing, circulate, screw it back together, run a free point, back off deeper, and continue until the fish is out or progress stops. If integral-joint tubing, such as PH or CS Hydril is present, the joint that is backed out should be run back into the hole to screw the tubing back together. Tubing-tool joints will swell after the string shot, unlike drill pipe or drill collars, on which a new joint or screw-in sub is run to screw them back together.

When performing a backoff, the pipe should first be tightened by applying a specific number of rounds of right-hand torque and then reciprocated while holding the torque. Pipe makeup torque should not be exceeded.

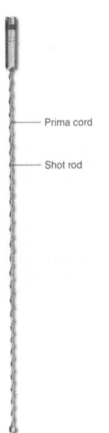

— Prima cord

— Shot rod

FIGURE 7-1

Shot rod with prima cord (courtesy of Baker Atlas).

Once the pipe is made up, left-hand torque is applied to the string. This torque must also be worked downhole by reciprocating the pipe as the torque is built up. This action distributes the torque throughout the string and ensures that there is left-hand torque acting on the point of backoff. A rule of thumb for determining the amount of left-hand torque is to use three-quarters of a round to one round for tubing and a half- to three-quarter-round for drill pipe for every 1,000 ft. of free pipe. Note that this is a base point. Some cases will call for more or less left-hand torque depending on hole conditions, such as angles and doglegs.

Ideally, the pipe at the backoff point should be in a neutral condition (under neither tension nor compression) before firing the string shot. Because this condition is very difficult to obtain, a slight tension (rather than compression) in the pipe should be favored. When the string shot is fired, a left-hand torque is held, and the calculated weight of the string is picked up. The concussion at the joint momentarily loosens the threads, and the left-hand torque on the pipe partially unscrews it. After repeating this process and unscrewing the joint completely with surface equipment, the freed pipe can be removed from the well.

When you are ordering a string shot, the service company will need to know the size and weight of pipe to be backed off, the approximate depth of the stuck point, the weight of the mud or fluid in the hole, and the temperature of the well. This information will dictate the strength of the charge needed as well as the type of fuse.

String shots may also be used for other purposes, including the following:

- Releasing stuck packers or fishing tools
- Removing corrosion from pipe
- Opening up perforations
- Jumping a collar or box on a tool joint
- Removing jet nozzles from drill bits to increase circulation
- Knocking drill pipe out of key seats in hard formations

OUTSIDE BACKOFF

Prior to running an outside string shot, the first question that should be addressed is the amount of open hole section from the previous casing shoe and the formations that are exposed. This technique is best suited for hard-rock application and is not advised to be run with adjacent shales or unconsolidated sands. As drilling wells has shifted to directional and long-reach horizontals, this technique is a very isolated case to near-vertical, hard-rock wellbores, and there is no other alternative method of releasing a stuck string.

An outside backoff is made by running string shots in the annulus to deliver concussions from outside the joint to loosen it. When pipe is plugged and cannot be cleaned out so a string shot can be run inside the pipe, an outside backoff is an alternative method of addressing the issue. In this procedure, a backoff is made internally as deeply as possible, and the free pipe is removed. A sub with a side opening is attached to the bottom of the string and screwed back into the fish downhole. An overshot can also be run on the bottom, but the outside diameter (OD) must be as small as possible to allow the wireline enough space to pass.

When the fish is screwed back together, or *latched*, the wireline and string shot are run inside the pipe and down to the side-door sub (also called an *outside backoff collar* or a *hillside sub*). There, it is guided into the annulus and lowered deeper. Torque is applied before running the wireline outside the sub. Once the backoff is completed, the pipe should be elevated slightly and rotated back to the right to untwist the line below the side-door sub. The string shot used for this operation has a small weight on the bottom of the line, with a 35-ft.-long prima cord wrapped around the wire. Because a collar locator is not run, and because most pipe lengths are between 28 and 32 in., this will ensure that the shot is positioned across a connection and the 35-ft. shot will hit the connection.

CHEMICAL CUT

In chemical cutting, the procedure employs an electrical wireline tool, a propellant, and a chemical reactant to burn a series of holes in the pipe, which weakens it so that it pulls apart easily. Introduced in the 1950s, chemical cutting was the exclusive, patented process of a single wireline company for many years. Today, most electric-wireline service companies offer it, and it is now the most widely used method of cutting pipe.

Wireline cuts are economical because rig time is minimized. Another significant advantage of chemical cutting is that there is no flare, burr, or swelling of the cut pipe (Figures 7-2A, B). No dressing of the cut is necessary to catch it on the outside, with an overshot, or on the inside, with a spear.

The body of the chemical cutting tool has a series of flow jets around its lower part. Electric current ignites the tool's propellant, which forces the chemical reactant (halogen fluoride or bromine trifluoride) through the jets under high pressure and at a high temperature, eroding the metal of the pipe. The tool also contains pressure-actuated slips to prevent vertical movement of the tool up the hole, which can foul the electric line.

The chemical cutting tool produces a series of perforations around the periphery of the pipe. The reaction of the chemical with the iron of the pipe produces harmless salts that do not damage adjacent casing and are rapidly dissipated in the well fluid.

The chemical cutter will not operate successfully in dry pipe. It requires at least 100 ft. of fluid above the tool when a cut is made. The fluid should be clean and contain no lost circulation material. The chemical cutting tool has operated successfully at a hydrostatic head pressure of 18,500 psi and 450°F. It is available for practically all sizes of tubing and drill pipe and most popular sizes of casing.

JET CUT

The jet cutter is a shaped charge run on an electric wireline. The modified parabolic face of the plastic explosive has a circular shape to make it conform to the pipe to be cut. The jet cutter is often used when abandoning a well, during salvage operations, or when a low fluid level, heavy mud, or cost considerations would preclude the use of a chemical cutter.

When a shaped-charge explosive is used to cut pipe, it flares the cut end of the pipe (Figures 7-3A, B). It is necessary to remove the flared pieces if the pipe is to be fished with an overshot from the outside. Usually this can be accomplished on the same trip as with the retrieving tool. A mill guide or a hollow mill container with an insert can be run on the bottom of an overshot and the flare or burr removed by

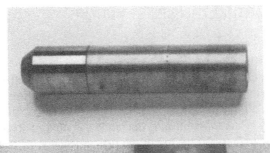

FIGURE 7-2

(A) Wireline chemical cutter; (B) tubing cut with the chemical cutter. (Both images courtesy of Baker Atlas.)

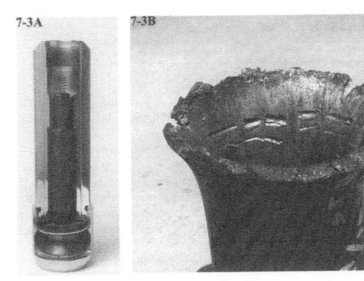

7-3A   7-3B

FIGURE 7-3

(A) Jet cutter; (B) tubing cut with the jet cutter. (Both images courtesy of Baker Atlas.)

rotation. Then the overshot can slip over the fish easily. There is a slight chance of damage to an adjacent string or casing if it is touching the pipe at the point where the cut is made.

Jet cutters are available for practically all sizes of tubing, drill pipe, and casing. Large jet cutters called *severing cutters* may be used if the fishing operation has stopped. They create such a large flare that it is almost impossible to dress off the top of the fish, especially in open holes. They are used to sever drill pipe and drill collars. The severing charge is placed across a tool joint and not in the tube of the pipe.

E-LINE MOTOR-DRIVEN CUTTERS

With the advancement in intervention techniques, coil, slick-line, and e-line, there are now options with coil, pressure-activated firing heads, slick-line timer-activated cutters, and e-line motor-driven cutters. Among these there are the Sondex cutter (General Electric, Fairfield, Connecticut), the Welltec cutter (Welltec, Alleroed, Denmark), and the BA/BHI Mechanical cutter.

All e-line cutters have a motor to drive the rotation, which acts to cut or displace the metal and force it apart. The cutter employs either a single-knife cutter or a three-blade system similar to a multistring cutter. The BA/BHI has a circular blade that is forced into the tube and separates the metal, acting very similar to a plumber's tube cutter

One benefit to e-line cutters is that they can be run on tractors, making internal cuts on pipe that is inclined greater than 60° possible. The other major advantage is that depth control with e-line is accurate with respect to couplings, connections, and packer bore (mandrel cutting), which makes it very convenient, with the common well profiles being drilled in areas where high angles and long course lengths can limit the use of other types of cutting methods.

One disadvantage of this technique, though, is its inability to cut heavier wall pipe, such as Hevi-Wate drill pipe. This limitation is caused as much by the expansion of the cutter as by the inside diameter (ID) restriction of the drill string.

MECHANICAL CUT

To run most fishing tools, the pipe is parted by wireline methods to minimize rig time. But if wireline tools are not available or practical, the pipe may be parted by running an inside mechanical cutter or a washover outside mechanical cutter on a workstring.

The internal cutter (Figure 7-4) is made on a mandrel and uses an *automatic bottom*. This allows the slips to be released and the tool to be set at any depth desired. Friction blocks or drag springs are fitted to the mandrel to furnish backup for this release operation. The mechanical inside cutter works by slowly rotating the tool to the right while slowly applying pressure, which feeds out the knives on tapered blocks. The knives then cut into the inside of the pipe. In practically all such cutting tools, springs are used in the feed mechanism to absorb accidental shocks to the workstring, which can cause the knives to gouge or break.

Fishing-tool operators will usually run a bumper jar above the inside cutter to control the pressure applied to it. The ends of the knives have a brass tip to prevent them from breaking when they come in contact with the pipe or casing wall. Inside mechanical cutters are available for most sizes of tubing and casing strings.

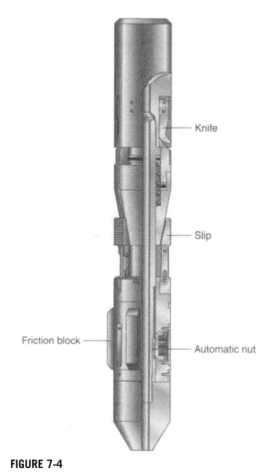

Knife

Slip

Friction block

Automatic nut

FIGURE 7-4

Mechanical internal cutter (courtesy of Baker Oil Tools).

INTERNAL HYDRAULIC CUTTER

The inside hydraulic cutter is designed to cut single strings of casing or tubing. The cutter has hydraulically activated knives for smooth and efficient cutting, an indicator that uses the pump pressure to signal when the cut is complete, and stabilizer slips to keep the cutter anchored in the casing. A piston forces the slips to anchor the tool firmly before the knives touch the casing.

When the cutter has been run to the desired cutting depth, rotation is initiated and circulation started. An increase in torque indicates that the casing or tubing is being cut. When the cut is complete, a "control dog" moves into a recess in the bottom nut of the cutter, causing a decrease in pump pressure. Straight pickup retracts the slips and knives. An inside hydraulic cutter should always be run with a float sub to prevent the cutter from setting because of hydrostatic pressure inside the workstring.

If the casing or fish is in an open hole, a pressure cutter, such as the multiple-string casing cutter, can be used (Figure 7-5). These cutters are designed for cutting more than one string, but they can be dressed with single-cut knives. Once the cutter reaches the appropriate depth, rotation starts and pump pressure pushes a piston down to make contact with the heels of the knives. This forces the knives into the casing. As more pressure is applied, the pipe is cut. When the cut is complete, pump pressure stops and the cutter is pulled out of the hole.

EXTERNAL CUTTER

The washover outside or external cutter (Figure 7-6) is used to go over the outside of a fish. This tool is ideal for tubing or pipe that is plugged up on the inside, preventing wireline tools from being run. The external cutter is run on the bottom of a washover string, and the cut is made from the outside. It is dressed to catch the type of tool joints or couplings on the fish. Pipe with couplings requires either a catcher assembly with spring fingers or flipper dog cages that catch below the coupling. Pipe that has couplings but also has upset joints can be caught with flipper dog–type or pawl-type catchers. These are made with slip surfaces cut on the end where they will engage the upsets.

Flush-joint pipe requires a hydraulically actuated catcher. Pump pressure against the sleeve restriction in the annular space actuates the external cutter's knives. You should begin slow rotation (at 15–25

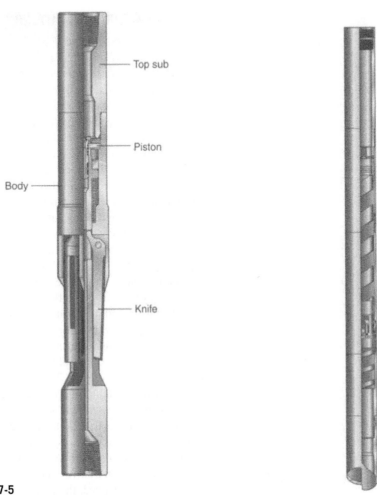

FIGURE 7-5

Multiple-string casing cutter (courtesy of Baker Oil Tools).

FIGURE 7-6

Washover outside cutter (courtesy of Baker Oil Tools).

rpm), and start to pump fluid slowly down the work string. This will begin feeding the knives to start the cut. The amount of pressure and number of gallons per minute required depend on the size of the cutter and the piston assembly being used. Use extreme caution to avoid surging the pump pressure when starting a cut. A decrease in torque will signal that the cut is complete. Once the cut is made, the cutter recovers the fish, and it is stripped out of the washpipe at the surface.

ABRASIVE FLUID CUTS

It has been known for a long time that adding an abrasive solid to a high-pressure fluid stream will cut steel. In the oil industry, this method has been employed successfully in many wellbores. The abrasive solid (typically angular sand) is jetted onto the steel with a specified standoff. Using this method, multiple strings can be cut with the caution of washing behind strings. The jetting fluid, water base, or oil base has to have only enough suspension properties to hold the solid in fluid until it is pumped.

Because hydraulic horsepower is being used, this method becomes somewhat restrictive to depth, pressure loss through string, and the type of work-string, just to mention a few of the considerations that must be taken into account. There is also the risk of washing out the work-string during the operation.

CONNECTION SPLIT SHOTS

A *split shot* is a perforation charge that has been elongated and focused to cut across a connection. As with the perforation shot, a split shot fires, and the explosive force inverts the shaped charge. As a result, all the force is concentrated in a vertical line across a connection and splits the connection (Figure 7-7). It is very common to have to work the connection apart after the shot fires. A successful shot will loosen the connection with minimal working in most cases.

FIGURE 7-7

Vertical split shot

CATCH TOOLS

OVERSHOTS

The overshot is the basic outside catch tool, and it is probably the most popular of all fishing tools. Overshots are used to externally engage, pack off, and pull a fish. They are designed with a helical groove in the bowl, and a grapple or slip made to fit is now almost universally used (Figure 8-1).

Most overshots consist of a bowl, top sub, guide, grapple or slip, control, pack-off, stop, and perhaps some additional accessory. The overshot bowl is turned with a taper on an internal helical spiral, and the grapple, which is turned with an identical spiral and taper, is fitted to it. Each grapple is turned with a slip or wickered surface inside so that a firm catch is ensured. Depending on the size of the catch for which it is designed, a grapple will be either a basket type (Figure 8-2A), for relatively small catches, or a spiral type (Figure 8-2B) for large fish, in relation to the outside diameter (OD) of the bowl.

The type of grapple used with the overshot is a function of fish size. Because a spiral grapple can appear flimsy, it sometimes causes concern about its strength. In practice, however, the spiral grapple makes a stronger assembly because it is flexible and distributes the load throughout the bowl. Note that most overshot failures occur because the tool is overstressed. The typical failure occurs when the bowl splits or swells because design limits have been exceeded.

Comparing the capacities of the two grapple designs illustrates the strength of a spiral grapple. A 7⅝-in., full-strength overshot fitted with a spiral grapple has a load capacity of 542,468 lbs. In the same configuration, the load capacity with a basket grapple is 479,044 lbs.

A cylindrical ring with a *tang,* or key, controls the location of the grapple within the bowl. This fits into a slot and prevents the grapple from turning but allows it to move up and down on the tapered surface. The grapple contracts as it is pulled down on the tapered surface, and it grips the fish more firmly as the pull is increased.

Controls may also be designed with a packoff or packer that seals around the fish and allows circulating fluid to be pumped through the fish. This can help to free a stuck fish.

Care must be exercised in fitting or "dressing" an overshot where a coupling or tool joint is to be caught. The enlarged tool joint or coupling section of the pipe to be caught must be positioned in the wickered area of the grapple. If it moves up and above this section, the overshot may rotate freely, and it becomes impossible to release. Stop rings of various designs are used to stop the enlarged catch in the proper grapple area. Some examples are doughnut-shaped rings placed in the bowl above the grapple, a spring-loaded packer or pack-off to seal inside a tubing coupling, and an internal shoulder at the top of the grapple itself.

Basket grapples with mill-control packers should always be run when fishing for drill pipe or tubing if the catch is small enough to accommodate them. Often there are burrs, snags, and splinters

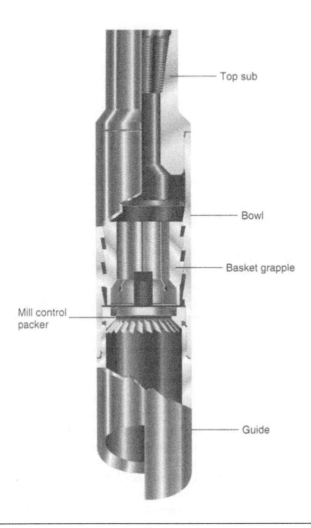

Top sub

Bowl

Basket grapple

Mill control
packer

Guide

FIGURE 8-1

Overshot dressed with a basket grapple.

(courtesy of National Oilwell)

on the pipe to be caught. The mill is sized so that these enlargements will be trimmed to the proper size. When the pipe has been "shot off," or parted in such a way as to heavily damage it, a mill extension or mill guide may need to be fitted to the overshot bowl so that the pipe can be milled on and caught in the same trip into the hole. These extensions or guides have tungsten carbide inside and can mill off a substantial amount of material so that the fish is trimmed to the grapple size (Figure 8-3).

Overshots are very versatile and may be fitted for almost any problem. Extensions such as washover pipe may be run from above so that pipe can be swallowed, and the overshot can catch the coupling or tool joint below. This is often done when the top joint of the fish is in such bad condition that it is not practical to pull on it.

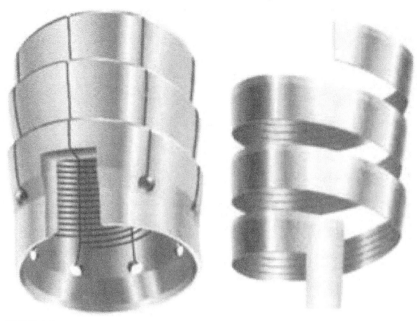

FIGURE 8-2

(A) Basket grapple; (B) spiral grapple.

(Both images courtesy of National Oilwell.)

There are various types of guides or shoes available to run below an overshot, including guides, mill guides, rotary guides, cut lip guides or will hooks. You can use the mule shoe or cut lip guide if the fish is difficult to catch because it is crooked or slightly out of gauge; use a wall hook guide if the top of the fish is laid over in a washout or cavity. (See Chapter 15 for more information.)

Short-catch overshots (Figure 8-4) are available in limited sizes to use when the exposed portion of the fish is too short to be caught with a conventional catch overshot. The wickered or catching portion of the grapples in short-catch overshots usually begin within 1 in. of the bottom of the bowl.

To properly engage an overshot on a fish, slowly rotate the overshot as it is lowered onto the fish (contrary to its name, an overshot should not be dropped over the fish). Circulation can be established to help clean the fish and to indicate when the overshot goes over the object being caught. Once this has been indicated by an increase in pump pressure, the pump should be stopped to avoid kicking the overshot off the fish. If jarring is to be done, it should be started with a light blow and gradually increased, as this tends to set the grapple on the fish. A hard impact may strip the grapple off the fish and cause the wickers to be dulled if the grapple is not properly engaged.

To release bowl-and-grapple overshots, the tapered surfaces of the bowl and the grapple must first be freed from each other. When pulling on a fish, these two surfaces are engaged and friction prevents release. Freeing or "shucking" the grapple is done by jarring down with the fishing string. A bumper jar is run just above the overshot for this purpose. After bumping down on the overshot, the grapple is freed, and the overshot is rotated to the right and released from the fish while taking a slight overpull.

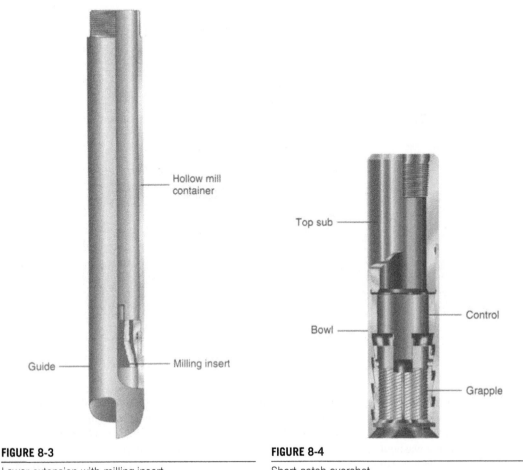

FIGURE 8-3

Lower extension with milling insert.

(courtesy of Baker Oil Tools)

FIGURE 8-4

Short-catch overshot.

(courtesy of National Oilwell)

DOUBLE BOWL

A double-bowl or combination overshot is essentially two single-bowl overshots run around each other. Dress the upper bowl with the smaller grapple (catch size). For example, dress the upper bowl to catch drill pipe or tubing and the lower bowl to catch the tool joint or tubing collar. The combination setup may be slightly more risky to run because of the longer length and higher number of fine-thread connections. Otherwise, a double bowl is operated similarly to the single-bowl setup.

TRIPLE BOWL

A triple bowl is similar to three single-bowl overshots run above each other. It gives you the ability to have three catch sizes. The smallest catch is in the top bowl and the largest catch is in the bottom bowl. It is seldom used because of the increased risk compared to using a single- or double-bowl setup.

WIDE CATCH OVERSHOT

The recent introduction (by National Oilwell) of the *Wide Catch Overshot (WCOS)* gives you a wider catch range for the undersized, oversized, or potentially unknown OD of the top of the fish. This tool has similar rugged design features and construction as the series 150 overshot (the industry standard), and it has the ability to interchange the bottom guide with the full range of existing components used with the series 150 overshot.

In action, the WCOS takes a positive grip over a large area of fish and is capable of withstanding heavy pulling, torsion, and jarring strains without damaging either the tools or the fish. This type of overshot has been designed to significantly increase the catch range of the OD of the fish compared to the standard overshot. This makes a successful fishing operation more likely to occur in a reduced number of trips, thus lessening the overall intervention costs for the operator. In addition to the large catch range, the WCOS has the ability to seal across very large extrusion gaps at both standard and high pressures and to provide full circulation through the fish if necessary.

New coarse threads have been introduced at the connection between the top sub and bowl to allow quick assembly while maximizing the torsion and tensile strength. An example of the catch range for the NOV WCOS for their 8.125 OD Series 150 overshot that is dressed with a nominal 7" grapple would be 6.719" to 7.031".

SPEARS

Spears (Figure 8-5) are used to catch the inside of a pipe or other tubular fish. Generally, a spear is employed only when an overshot is not suitable. The spear has a small internal bore that limits running some tools and instruments through it, and it is used for such tasks as cutting, free-pointing, and in some cases, backing off. Note that spears are more difficult than overshots to pack off or seal between the fish and the workstring.

However, spears are more useful than overshots for fishing jobs such as pulling liners and packers, picking up parted or stuck casing, or fishing pipe that is enlarged because of explosive shots, fatigue, or splintering. Due to their design, which features a small bore in the mandrel, spears are usually very strong. For example, one manufacturer's spear for picking up 5½-in. casing with a 4½-in. drill pipe has a capacity of 628,000 lbs. An overshot made for the same catch would have a capacity of 580,000 lbs. Note that either tool in this size is adequate because the yield strength of 4½-in., 16.60-lb. Grade S drill pipe is 595,000 lbs. and Grade E is 330,000 lbs.

The spear is a versatile tool. It can be run in the string above an internal cutting tool or in combination with other tools, saving a trip into

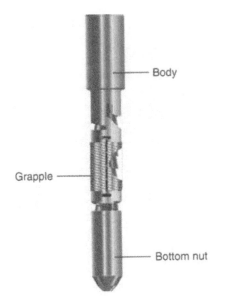

Body

Grapple

Bottom nut

FIGURE 8-5

Releasing spear.

(courtesy of National Oilwell)

the hole with the workstring. Milling tools may be run below the spear to open up the pipe so that it can enter and catch.

The most commonly used spears are built on the same principles as overshots. They are designed with a tapered helix on the mandrel (similar to the tapered helix turned inside the bowl of an overshot) and a matching surface on the inside of the grapple. The slip, or gripping surface, of the grapple is on the outside surface of the spear so that it will catch and grip the inside of the pipe being fished.

To release a spear, it is rotated to the right. If the grapple is frozen to the mandrel, it may be necessary to bump down to free, or "shuck," the grapple. A bumper jar can be run just above the spear to jar down and free the grapple.

To pack off the fish when catching with a spear, place an extension below the spear with the appropriate packoff cups facing down (Figure 8-6). Typically, these are protected by a steel guide, which helps the packoff cups to enter the pipe without damage.

A stop sub is run above the spear to space it properly in the fish and to aid in releasing the spear. It also allows weight to be set down on the spear to cock the jars (Figure 8-7). The grapple should be

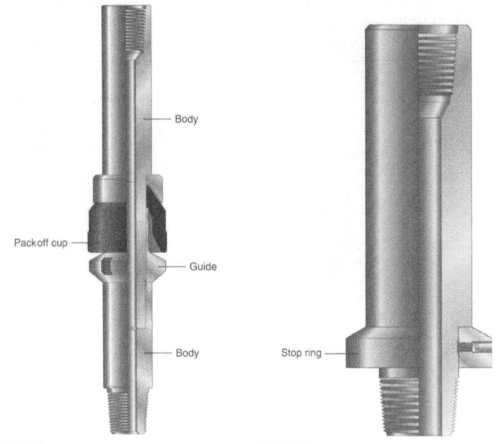

Body

Packoff cup

Guide

Body

Stop ring

FIGURE 8-6

Spear packoff.

(courtesy of Baker Oil Tools)

FIGURE 8-7

Spear stop sub.

(courtesy of Baker Oil Tools)

placed far enough into the fish to secure a good grip, but if it is too far below the top of the pipe, it may swell the pipe if great force is applied. Releasing the spear then becomes a problem. Ordinarily, the stop sub is placed 1 to 2 ft. above the catching surface of the spear. However, extensions may be used to place it lower in the fish if the top is splintered or swelled.

There are other spear designs as well. Most of these are built on the tapered-wedge principle. J releases, automatic bottoms, and split nuts and cams are used to set and release the slips or grapple. These types of spears vary with the different fishing-tool companies. The overshots and spears discussed in this chapter are widely used by all companies. Hydraulically operated overshots and spears are made mostly for use in thru-tubing fishing (discussed further in Chapter 24).

IMPACT TOOLS

PRINCIPLES OF JARRING

Simply stated, a jar is a delay mechanism. When installed in a fishing string and connected to a stuck object, a jar allows the operator the time required to pull and stretch the string. Using a hydraulic or oil jar, the pipe is pulled to a determined prejarring load while stretching the fishing string. The stretched drill string is held in tension, storing energy like a taut rubber band.

The hydraulic fluid inside the jar meters from one side of a piston to the other side within a pressure chamber, allowing the piston to travel slowly to an area with a larger bore size, where resistance to opening no longer exists.

When the detent is released, the jar opens to its full, stroked-out length, the energy stored in the stretched pipe is released and the string returns to its original length, producing kinetic energy that makes the jar's piston hit the striking surface with a sharp upward blow.

As high pressure builds up inside the jar during the prejarring load, pulling limits must not be exceeded or else the jar body could rupture and cause the tool to part. Check the manufacturer's specifications before jarring. After the jar fires, it is possible to handle as great a load as the fishing string will safely permit.

If the jars are worked too fast and with too heavy a load, the hydraulic oil will become hot, lose viscosity, and fire prematurely. As a result, the operator is not given sufficient time to stretch the pipe.

SAFETY

All care should be taken to prevent accidents during fishing operations, and jarring can be hazardous. The shock that is delivered to the fish can affect the rig and its components. Things can shake loose and fall. It is best practice to have a prejob safety meeting and a prejarring inspection to make sure that all equipment and fixtures in the derrick are secure. Any nonessential personnel should leave the drill floor, and a postjarring operation should be conducted. Each operator and drilling contractor will have individual policies about jarring, and they should be reviewed and followed.

JARRING ON STUCK TOOLS

Early cable-tool drillers used link jars for both fishing and drilling. Today, jars fall into two categories: *fishing* and *drilling*. Jars in each of these categories can be further distinguished by either their *hydraulic*

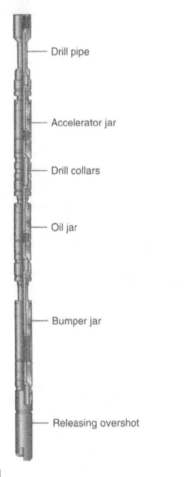

Drill pipe

Accelerator jar

Drill collars

Oil jar

Bumper jar

Releasing overshot

FIGURE 9-1

Typical jarring bottomhole assembly.

(courtesy of Baker Oil Tools)

or *mechanical* operation. While both fishing jars and drilling jars work on the same principle, they are usually built quite differently.

Most jarring strings (Figure 9-1) for fishing consist of an oil jar (sometimes called a *hydraulic jar*) and a bumper jar (also called a *bumper sub*), along with the necessary drill collars to add weight. Also, an accelerator (also called an *intensifier* or *booster jar*) may be added to the string.

The oil jar strikes an upward blow to the fish, and the bumper jar strikes a downward blow. An accelerator can be included with the jarring string to provide additional stored energy, which helps to speed up the drill collars when they are released by the oil jars. It also provides free travel that compensates for the travel of the oil-jar mandrel. This travel compensation prevents the workstring from being pushed up the hole, which absorbs the energy of the impact through friction.

JARRING STRINGS

A complete jarring string (shown previously in Figure 9-1) consists of an appropriate catching tool or screw-in sub, a bumper jar or sub, an oil jar, the desired drill collars, an accelerator jar, and a workstring. Each tool in the string performs a specific function, and it is critical to run them in the proper sequence. The overshot or spear on the bottom catches or engages the fish. The bumper jar moves downward, either to dislodge the fish or to help release the catching tool. The oil jar imparts the upward blow, the drill collars furnish the weight necessary for a good impact, and the accelerator speeds up the jarring movement and compensates for the travel in the oil jar, while saving the pipe from compressive stresses.

The weight of the drill collars run with jarring strings varies according to the size of the jar used and is also influenced by the depth of the fish, the fluid in the well, the strength of the workstring, and the amount of fish stuck in the wellbore. Manufacturers of oil jars and booster jars make recommendations for a range of weights to be run with their tools. Today, most manufacturers have computerized jar-placement programs (see the section "Jar Placement Programs," later in this chapter) that give the operator the exact location and number of drill collars to be run according to the information entered.

When accelerator jars are run, it is important not to use excessive drill collar weight. Too much weight tends to overload the accelerator and hinder, rather than help, the action. Do not run drill collars or heavyweight pipe above the accelerator. This procedure causes the string to elongate when the jars trip and may deliver an impact to the accelerator, which can damage the tool. If drill collars or heavy-weight pipe in the string must be included for length, they may be placed farther up the hole, where they will not affect the jarring operation.

To activate the oil jar, pull up to a predetermined distance above the weight of the string at the jar. Hold this position while the oil jar bleeds off and the blow is delivered. This jarring weight may be adjusted to any amount, so long as it stays within the strength of the tools and pipe run. Do not pull the jar past its recommended strength, as this will damage it. Ordinarily, jarring is started at a low impact and gradually increased as necessary. This ability to vary the impact is the most significant advantage of hydraulic or oil jars over mechanical jars. When running an overshot or other catching tool, it is particularly desirable to start jarring at a low impact and increase as necessary. This sets the grapple and allows it to "bite" into the fish. If a heavy blow is struck first, it can cause the grapple to strip off the fish, leaving the grapple dull and unable to catch again.

In jarring up, the bumper jar has no function and merely acts as an extension or slip joint. When jarring down is desirable, the oil jars should be closed and the stroke of the bumper jar used for the downward impact. This precaution is necessary to prevent jarring down on the oil-jar packing, which would cushion the downward impact and possibly damage the packing.

Jars are always redressed after each use, even if they did not strike a blow. They are disassembled and inspected, and new seals and oil are installed. Then they are tested on a pull rack for resistance to pull according to size.

Oil jars are never run in pairs because they do not trip at the same time, and one would affect the packing and seals of the other. This would damage the packing and hinder good jarring action.

Jars should always be replaced when making a trip for a purpose other than to change them. There is no way to determine the remaining useful life of a jar that has been operated. Replacement is the best insurance.

The addition of drill collars or heavyweight drill pipe above the oil jar was formerly thought to create an increase in impact, but it was not always possible to move this mass fast enough to be effective. The impulse or duration of the impact is also important, particularly for long sections of stuck fish such as differentially stuck drill collars. However, data and experience strongly suggest that the drill-collar weights recommended by the manufacturer should not be exceeded. If possible, applied weight should be kept in the low range.

JAR PLACEMENT PROGRAMS

The effectiveness of a fishing jar depends not only on its design, but also on its placement in the fishing string, fishing string design, hole conditions, and description of fish.

Because the analysis of the mechanics of jarring is complex, a computer program is needed to precisely determine optimum jar placement. Sometimes moving the jar only a few feet up or down the fishing string can drastically change how the stress waves add up to create a jarring force on the stuck point. It is always recommended to contact your local jar company to have a jar placement program run for the optimum chance of success in recovering a fish.

BUMPER JARS

A bumper jar (Figure 9-2) is a mechanical slip joint. The simplest models have exposed mandrels when open. In other designs, the mandrel splines are enclosed and lubricated. The bumper jar is used almost exclusively as a downward-impact tool. The bumper jar releases the weight of the drill collars that it carries, which causes a heavy impact as it closes. In addition to delivering impact blows to the fish, bumper jars are used above catching tools such as overshots and spears. If the tapered grapples or slips of the tools become stuck on the mandrels or in the bowls, they may be jarred down off the tapers by bumper jars. This releases the tool from the fish.

Fishing-tool operators frequently use bumper jars in a string of fishing or cutting tools so that constant weight may be applied to a tool such as a cutter. By operating within the stroke of the bumper jar, only the weight below that point is applied to the tools. One example is a cutter in a deviated hole; the weight run below the bumper jar is exerted on the knives, but excessive weight from the workstring is avoided. The bumper jar is sometimes called "the fisherman's eyes." The free travel in the bumper jar can easily be seen on the weight indicator, which helps the fishing operator apply the correct weight to tools such as dress-off mills or taps when screwing them into a fish.

FIGURE 9-2

Fishing bumper jar.

(courtesy of National Oilwell)

OIL JARS

An oil jar (Figure 9-3) consists of a mandrel and piston operating within a hydraulic cylinder. When the oil jar is closed, the piston is in the down position, where it has a very tight fit and restricted movement within the cylinder. The piston is fitted with specialized packing that slows the passage of oil from

FIGURE 9-3

Hydraulic oil jar.

the upper chamber to the lower chamber of the cylinder when the mandrel is pulled by picking up on the workstring. About halfway through the stroke, the piston reaches an enlarged section of the cylinder, which frees its movement. The piston then moves up very quickly and strikes the mandrel body. The intensity of the impact can be varied by the amount of strain on the workstring. This variable impact is the main advantage of the oil jar over the mechanical jar for fishing.

Some oil-jar designs have check or bypass valves that allow fluid to quickly transfer to the chamber above the piston when cocking or reloading. However, many oil jars in the field do not have this feature. In older designs, fluid must transfer through gaps in seals and rings. If weight is applied too rapidly to close the jars, fluid will pass around the seals and damage them, shortening the life of the jars. Use caution and slack-off weight slowly when reloading jars to prevent this damage.

Some oil jars also incorporate a floating piston that effectively transfers the pressure of the hydrostatic head to the jar fluid. Oil jars are very effective in freeing stuck fish because the energy stored in the stretched drill pipe or tubing is converted to an impact force, which can easily be varied according to the pull exerted on the workstring. Most oil jars are effective up to 350 °F, but they also can be used with special heat-resistant oil that will tolerate higher temperatures.

BIDIRECTIONAL HYDRAULIC FISHNG JAR

In horizontal wellbores, it can be difficult to make an impact on the fish in either an upward or downward motion. By running a bidirectional hydraulic fishing jar (Figure 9-4), you can achieve a dynamic, double-acting impact in an upward or downward movement. Most horizontal wells being drilled today have a top drive system, and due to the nature of such a system, they are slow moving. As a result, it is difficult to make a downward impact due to the wellbore geometry. Should there be a need to release from an overshot, this setup will achieve the downward motion needed to create a grapple bite to release the fish.

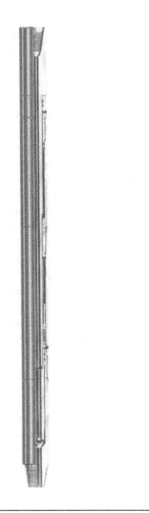

FIGURE 9-4

Bidirectional fishing jar.

(courtesy of Logan Oil Tools)

JAR INTENSIFIER OR ACCELERATOR

The intensifier or accelerator, also called a *booster jar* (Figure 9-5), is an accessory run in the jarring string. When run above drill collars, the impact delivered to the fish is increased, and most of the shock is relieved from the workstring and rig. The intensifier is essentially a fluid spring composed of a cylinder filled with a compressible fluid, which is usually an inert gas or silicone. When the workstring is pulled, a piston in the cylinder compresses the fluid and stores energy. When the oil jar trips, this energy is released, and it speeds the movement of the drill collars up the hole to deliver a heavier blow.

The free travel that is made available when the accelerator is pulled open also relieves the workstring of the majority of the rebound impact, which is damaging to tools and tool joints. The free travel in the accelerator compensates for the free stroke of the oil jar. Ordinarily, without an accelerator in the string, the workstring is stretched, and when the oil jar trips, the pipe is released to move up the hole. Much of the stored energy is absorbed by friction in the wellbore. This is apparent by movement at the surface, causing the elevator, traveling block, and even the derrick to shake. An accelerator in the string eliminates this movement.

Isolating this sudden compressive force from the workstring is one significant advantage of an accelerator. Because the jar's impact is increased by the higher speed with which the drill collars move up to strike a blow, less weight or mass is required. Manufacturers offer recommended weights of drill collars to be run with each accelerator jar.

During the past decade, one jar manufacturer presented the option to use dual accelerators in their software to further enhance the jar's capability. In some regions, this has become a standard jarring string configuration. There are several benefits to such a configuration, the first of which is higher impact and impulse forces for the equivalent overpull. The stored energy of the workstring, as well as the additional free stroke when the jar fires, is transmitted directly to the fish, maximizing the effectiveness of the jarring string.

This arrangement is particularly beneficial with shallow jarring because it reduces the shock load to surface equipment. When the shock load from shallow jarring is considered along with the top drives in use today, this reduction in the load on the top drive is substantial. This assembly is effective in

FIGURE 9-5

Intensifier jar.

(courtesy of National Oilwell)

horizontal wells because it places dual accelerators in wells where the inclination is less than 65°.

As with all newer techniques, there are usually drawbacks, and this is no exception. As mentioned earlier in this chapter, staging the jarring loads in steps is important and should be performed regardless of the jarring string configuration. With the dual accelerator configuration, this becomes critical as the increased loads, depending on the well and the fish, may make the equipment fail at maximum jarring. Therefore, a rule of thumb when using dual accelerators is not to exceed 90 percent of the manufacturer's recommended maximum jarring load.

AGITATOR TOOLS

In recent years, a new form of releasing stuck pipe has been developed. Although it has not yet been fully proven with field results, it has been used in a number of successful recoveries, and as the technology evolves, improvements are evident. Agitator tools are used in drilling to achieve improved rates of penetration, in coiled tubing applications for getting tools into longer horizontal wellbores, and in fishing operations to support the recovery of stuck pipe, bottom-hole assemblies (BHAs), and point stuck equipment, such as packers.

THE FISHING AGITATION TOOL (FAT)

The Fishing Agitator has been introduced to improve fishing BHA efficiency by providing a new engine to conventional stroking tools. The Fishing Agitator is a down-hole component that turns hydraulic power (fluid in the string) into mechanical energy in the form of axial oscillation. This new technology enhances the performance of conventional fishing operations by changing the impact frequency enormously, from 1 blow per minute (with the Fishing Jar) to 1,200 blows per minute (with the Fishing Agitator). The axial vibration generates a higher cumulative energy per minute than does the jarring mechanism.

The Fishing Agitator acts by breaking static friction applied by the formation on the stuck pipe and turning it into dynamic friction at lower friction coefficients. When tension is applied to the fishing string while engaged to the fish and pumping through the Fishing Agitator, microblows at a 20-Hz frequency work to slide the fish away from the sticking area. An agitator assembly can be utilized as a standalone product with a catch tool, or it can be run with a jarring assembly to further enhance the recovery of a stuck object.

SURFACE JARS

Occasionally, a drill string can become stuck near the surface, primarily in keyseats or at the bottom of the surface pipe. To free this pipe, it is necessary to strike a heavy blow downward, as jarring upward would only cause the fish to become more stuck. Early drillers fashioned a *driving joint* from an old kelly or joint of pipe and a sleeve or large pipe outside that would slide on the inner body. Two flanges, one on each member, were used as the striking faces. This was made up in the string at the surface, and then the outer body was picked up with the catline and dropped. This delivered a significant impact against the flange in the string and frequently freed it, thus saving an expensive fishing job.

In modern fishing, the old driving joint has been replaced with a surface bumper jar (Figure 9-6), which also delivers a heavy downward blow. By adjusting the friction slip in the jar, the impact may be increased or decreased as required. The jar is made up in the string at the surface, and the friction slips and the control ring are adjusted to the desired tripping pull. When a straight upward pull is exerted on the jar, the friction slip rubs the enclosed friction mandrel and arrests upward movement while the drill pipe is being stretched. When the upward pull reaches the preset tripping weight, the friction mandrel is pulled through the friction slip. The resulting downward surge of the pipe as it returns to its normal length causes a sudden separation of the main mandrel and bowl assemblies. Then, they are free to move apart for the length of its 48-in. stroke and drive the weight of the free pipe against the stuck point.

As in all jarring operations, light blows should be used at first with a surface bumper jar. If light jarring fails, the tool may be adjusted for heavier impacts. The tripping weight should never be set higher than the weight of the free pipe between the jar and the stuck point. If the jar is set higher than this weight, it becomes necessary to pull on the pipe at the stuck point, which will usually cause it to stick more. Only hit a surface jar a few times, as excessive jarring with a surface jar can break the drill string. The pipe will start to bend in a washed-out area and can break at this point eventually.

Occasionally, fishing tools that operate on tapers may become frozen. The wickers on grap-

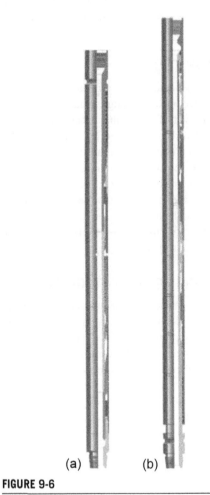

(a)   (b)

FIGURE 9-6

Surface bumper jar.

(courtesy of National Oilwell)

pling tools may bite into the fish so deeply that the tools cannot be released in the normal manner. A bumper jar with a downward impact is very effective in freeing the grapple so the tool may be released. The surface bumper jar is also used for this purpose.

DRILLING JARS

In most drilling conditions, it is economical to run jars in the drill string so they are readily available if the string becomes stuck. As mentioned previously, these jars come in two designs.

The hydraulic or drilling oil jar (Figure 9-7A) is essentially the same design as the fishing oil jar, except that it is made much heavier and stronger to sustain several hundred hours of drilling. The down jar is essentially a mechanical or hydraulic detent design (Figure 9-7B) using friction slips and mandrels. Drilling jars are much longer than conventional fishing jars.

Drilling jars should ordinarily be run in tension above the neutral point of the string. If placed in the transition zone, the jars would be subjected to unusual flexing, which could cause premature failure. Run above the majority of drill collars, the jars are readily available if the bit or drill collars stick. Several drill collars or heavy-weight pipe may be run above the jars to increase their impact with additional mass. Manufacturers furnish instructions for each specific design.

Impact forces exerted by jars are expressed in a unit of measure called "pounds jarring." This value is purely theoretical and is derived from the amount pulled on the workstring above its normal weight. True impact force is affected by many variables such as mud weight, friction in the hole, drill collar mass or weight, and the stroke of the jars. Certain theoretical calculations are used primarily to prevent jarring with too much pull, weight, or both. Remember that the purpose of a jar is to move the fish without hitting it so hard that it splits.

FIGURE 9-7

(A) Hydraulic up-jar section; (B) Mechanical down-jar section.

(courtesy of Logan Oil Tools)

WASHOVER OPERATIONS

Washover pipe (or *washpipe,* as it is commonly called), is large pipe used to drill out, wash out, and circulate out cement, fill, formation, or other debris that is causing the fish to stick. Proper size selection in washover operations is critical. The washpipe's inside diameter (ID) must be large enough to go over the fish, with clearance for circulation. The outside diameter (OD) must allow rotation in the hole or casing. To avoid sticking the washpipe, its annular clearance must be sufficient for circulation and prevention of overtorquing (specifications are given in Figure 10-1).

A typical washpipe string is composed of a top bushing or safety joint, the desired number of pipe joints, and a rotary shoe on the bottom that is suitable for the material needing to be cut. The top bushing adapts washpipe threads to a tool joint to fit the workstring. Safety joints are sometimes run in place of the top bushing.

Two types of safety joints are used with a washover string. One, a standard safety joint, will be unscrewed at 7 to 9 rounds to the left if the washpipe becomes stuck. The other is a washover/backoff-type safety joint. Unscrewing this safety joint takes 15 to 17 rounds to the left. Without the safety joint, the washpipe could unscrew completely while trying to back off of the fish. If this happens, the fish can be retrieved, but the washpipe will be left in the well.

Washpipe is usually made of heavy-wall, N-80-grade casing cut into 30- to 33-in. (Range 2) lengths for handling ease, with special threads for good torquing and strength characteristics. A washover operation is actually a drilling procedure, so the pipe is subjected to high torque. Shoulders like those used on tool joints are usually part of the washpipe thread design and have the strength to withstand a high torque.

For maximum clearance, washpipe joints are usually flush inside and outside. This also allows the inside casing to be run for workover operations. Where differential sticking is a problem in open holes, pipe with external upsets or couplings, such as the popular X-line upset-connection joint, is used.

ROTARY SHOES

The rotary shoe (Figure 10-2) is run on the bottom of the washpipe string. Various types are available for specific tasks. Tooth-type shoes are typically used when cutting fill, formation, or cement. The teeth on these shoes have straight leading edges, and their surfaces are dressed with a material such as tube borium to prevent excessive wear and to resist erosion by the circulating fluids.

Shoes for cutting steel such as tool joints, tubing, or junk are dressed with tungsten carbide in an appropriate configuration. Shoe designs must allow sufficient circulation to cool the carbide and wash away cuttings. If the job is inside casing, cutting carbide must not contact the casing wall, as this

Connection	OD		Weight Plain End		Inside Diameter		Wall Thickness		Upset Diameter		Drift Diameter		Washover Size Rec.		Max.	
in.	in.	mm	lbs/ft	kg/m	in.	mm	in.	mm	in.	mm	in.	mm	in.	mm	in.	mm
TSWP	3-3/8	85.73	10.02	14.89	2.764	70.21	0.305	7.75	-	-	2.639	67.03	2-1/2	63.05	2-5/8	66.68
TSWP	3-1/2	88.9	12.31	18.3	2.764	70.21	0.368	9.35	-	-	2.639	67.03	2-1/2	63.5	2-5/8	66.68
TSWP			8.81	13.09	2.992	76.0	0.254	6.45	-	-	2.867	72.82	2.69	68.26	2-7/8	73.03
TSWP	3-5/8	92.08	7.06	10.49	3.240	82.3	0.192	4.88	-	-	3.115	79.12	3.00	76.2	3-1/8	79.38
TSWP	3-3/4	95.25	9.55	14.19	3.238	82.25	0.256	6.5	-	-	3.113	79.07	3.00	76.2	3-1/8	79.38
TSWP			10.46	15.55	3.185	80.9	0.283	7.19	-	-	3.06	77.72	3.00	76.2	3-1/8	79.38
TSWP	3-13/16	96.84	11.7	17.39	3.187	80.95	0.313	7.95	-	-	3.062	77.77	3-1/16	77.79	3-1/8	79.38
TSWP	4	101.6	12.93	19.22	3.340	84.84	1.33	8.38	-	-	3.215	81.66	3-1/2	88.9	3-1/4	82.55
TSWP	4-3/8	111.13	12.02	17.87	3826	97.18	0.275	6.99	-	-	3.701	94.01	3-1/2	88.9	3-3/4	95.25
TSWP			13.8	20.18	3.749	95.22	0.313	1.95	-	-	3.624	92.05	3-1/2	88.9	3-5/8	92.08
TSS	4-1/2	114.3	11.3	16.87	4.000	101.6	0.25	6.35	-	-	3.795	98.43	3-3/4	95.25	3-7/8	98.43
TSWP			13.04	19.38	3.920	99.57	1.29	7.37	-	-	3.701	96.39	3-5/8	82.08	3-3/4	95.25
TSWP			14.98	22.26	3.826	97.18	0.337	8.56	-	-	3.701	94.01	3-1/2	88.9	3-11/16	93.66
TSS	4-3/4	120.65	17.2	26.04	4.000	101.6	0.375	9.3	-	-	3.875	98.43	3-3/4	9.25	3-7/8	98.43
TSWP	4-7/8	123.83	11.57	17.2	4.408	112.0	0.233	5.92	-	-	4.283	108.8	4.00	104.6	4-1/8	104.8
TSWP	5	127.0	14.87	22.1	4.408	112.0	0.296	7.52	-	-	4.283	108.8	4.00	101.6	4-1/8	104.8
X-LINE			15.00	22.29	4.375◆	111.1			5.36	136.1	4.151	105.4			4-1/4	108.0
X-LINE			18.00	26.75	4.250◆	108.0			-	-					4-1/8	104.8
TSWP			17.93	26.65	4.276	108.6	0.362	9.19	-	-						
TSS	5-3/8	136.3	20.2	29.76	4.625	117.5	0.375	9.53	-	-	4.5	114.3	4-1/4	107.95	4-1/2	114.3
TSWP	5-1/2	139.7	16.87	2.07	4.892	124.3	0.304	7.72	-	-	4.767	121.1	4-5/8	117.48	4-3/4	120.7
X-LINE			17.00	25.27	4.875◆	123.8			586	148.8						
TSWP			19.81	29.44	4.778	11.4	0.361	9.17	-	-	4.653	118.2	4-1/2	114.3	4-5/8	117.5
TSWP	5-3/4	146.05	18.18	7.02	5.124	130.2	0.313	7.95	-	-	4.999	127.0	4-7/8	123.83	5	127.0
TSS			21.53	32	5	127.0	0.375	9.53	-	-	4.875	123.8	4-3/4	120.65	4-7/8	123.8
TSS	6	152.4	15.35	22.81	5.5	139.7	0.25	6.35	-	-	5.375	136.5	5-1/4	133.35	5-3/8	136.5
TSWP			19.64	29.19	5.352	135.9	0.324	8.23	-	-	5.227	132.8	5-1/8	130.18	5-1/4	133.4
TSWP			22.81	33.9	5.24	133.1	0.38	9.65	-	-	5.115	129.9	5.00	127.0	5-1/8	130.2
TSWP	6-3/8	161.93	24.03	35.72	5.625	142.9	0.375	9.53	-	-	5.5	139.7	5-3/8	136.53	5-3/8	139.7
TSWP	6-5/8	168.28	23.58	35.05	5.921	150.4	0.352	8.94	-	-	5.796	147.2	5-5/8	142.88	5-1/2	146.1
X-LINE			24.00	35.67	5.91	150.4			7	177.8						
TSWP	7	177.8	25.66	38.14	6.276	159.4	0.362	9.19	-	-	6.151	156.2	6.00	152.4	6-1/8	155.6
X-LINE			26.00	38.64	6.276◆	159.4			7.39	187.7						

Type	Size															
TSWP	7-1/4	184 15	23.19	34.47	6.624	168.3	0.313	7.95	-		6.499	165.1	6-3/8	161.96	6-1/2	165.1
TSWP	7-3/8	187.3	28.04	41.68	6.63	168.3	0.38	9.5		0	6.5	165	6-3/8	161.96	6-1/2	165.1
TSWP	7-5/8	193.7	25.56	37.99	6.97	177	0.33	8.3	-		6.84	174	6-3/4	171.5	6-7/8	174.6
TSWP			29.04	43.16	6.88	174.6	0.38	9.5		0	6.75	172	6-5/8	168.3	6-3/4	171.5
TSWP			33.04	49.11	6.77	171.8	0.43	10.9			6.64	169	6-1/2	165.1	6-5/8	168.3
X-LINE			29.70	44.14	6.843◆	173.8◆	0.38	9.5	8.01	204	6.75	17.2	6-9/16	166.7	6-11/16	169.9
TSWP	8	203.2	30.54	45.39	7.25	184.2	0.38	9.5	-	0	7.13	181	7	177.8	7-1/8	181
TSWP	8-1/8	206.4	31.04	46.13	7.38	187.3	0.38	9.5	-	0	7.25	184	7-1/8	181	7-1/4	184.2
TSWP			35.92	53.39	7.25	184.2	0.44	11.1	-	0	4.13	181	7	177.8	7-1/8	181
TSWP			38.42	57.1	7.19	182.5	0.47	11.9			7.06	179	6-15/16	176.2	7-1/16	179.4
TSWP	8-3/8	212.7	33.95	50.46	7.58	192.5	0.4	10.1	-	0	7.45	189	7-1/4	184.2	7-3/8	187.3
TSS			37.09	55.13	7.5	190.5	0.44	11.1			7.38	187				
TSWP	8-5/8	219.1	31.1	46.22	7.92	201.2	0.35	8.9	-	0	7.8	198	7-9/16	192.1	7-11/16	195.3
X-LINE			36	53.51	7.813◆	198.5◆	0.4	10.2	9.12	232	7.7	196	7-1/2	190.5	7-5/8	193.7
TSWP			39.29	58.4	7.73	196.2	0.45	11.4	-	0	7.6	193			7-11/16	195.3
TSWP	9	228.6	38.92	57.85	8.150	207	0.43	10.8	-	0	7.99	203	7-7/8	200	8	203.2
TSWP	9-5/8	244.5	38.94	57.88	8.84	224.4	0.4	10	-	0	8.68	220	8-1/2	215.9	8-3/8	212.7
TSWP			42.7	63.46	8.76	222.4	0.44	11	-	0	8.6	218	8-1/4	209.6	8-1/2	215.9
X-LINE			43.5	64.65	8.67	220.1	0.47	12	10.1	257						
TSWP			46.14	68.58	8.68	220.5	0.4	12		0	8.53	217				
X-LINE	10-3/4	273.1	44.22	65.72	9.95	252.7	0.4	10.2		0	9.79	249	9-1/2	241.3	9-3/4	247.7
TSWP			49.5	73.57	9.85	250.2	0.45	11.4		0	9.69	246	9-3/8	238.1	9-5/8	244.5
TSWP	11-3/4	298.5	54.21	80.57	9.76	247.9	0.5	12.6		0	9.6	244	9-1/4	235	9-1/2	241.3
TSWP			52.57	78.13	10.9	276.4	044	11			10.7	272	10-1/8	257.2	10-5/8	269.9
TSWP	12-3/4	298.5	58.81	87.41	10.8	273.6	0.49	12.4			10.6	270	10	254	10-1/2	266.7
TSWP	13-3/4	339.7	49.56	73.66	12	304.8	0.38	9.5	13.5	343	11.8	301	11	279.4	11-1/2	292.1
TSWP	16	406.4	66.11	98.26	12.4	315.3	0.48	12.2	13.75	349	12.3	311	11-1/2	292.1	12	304.8
TSWP			81.97	121.68	15	381.3	0.5	12.6	16.75	426	14.8	377	14-1/4	362	14-3/4	374.6

All strengths maximum value - apply safety factor of two (2) to the joint tensile yield strength.

* = Recommended make-up torque = 25% of maximum make-up torque - does not apply to X-Line Connections.

■ = Ratio of the joint tensile yield strength to the pipe tensile yield strength.

◆ = The internal upset has been reduced.

● = N-80 Pipe.

◊ = J-55 Material.

FIGURE 10-1

Washpipe data specification guide.

(courtesy of Baker Oil Tools)

FIGURE 10-2

Washpipe rotary shoe.

(courtesy of Baker Oil Tools)

will damage it. To prevent this, smooth brass is sometimes applied to the shoe's outside diameter as a bushing that reduces friction and prevents damage. Tungsten carbide is applied to the bottom of the shoe, and if possible, to the ID. A small carbide shoulder inside the shoe improves the chances of retrieving some of or all the fish. This can save a trip with another tool to recover the washed-over fish.

Because washpipe is large, stiff, and smooth, length of the washpipe string is a critical factor in preventing sticking. Length should be determined after careful consideration of hole conditions. Note that deviated holes or accidentally crooked holes limit the safe length of a washpipe string. The following discussions of two actual jobs demonstrate the extremes in washpipe-string lengths.

EXAMPLE

In one job, drill pipe in a vertical well was stuck from a depth of 330 ft. (the bottom of the surface pipe) to 8,487 ft. (the depth of the bit). The cause of the sticking was a poor mud system. The mud had to be circulated out and replaced, conditioning the hole in the process. On the last washover, 1,218 ft. of washpipe was run. This is unusual, but under the circumstances, the decision was correct and the job was successfully completed.

In this situation, equipment costs, as well as the cost of drilling the hole, the cost of the surface pipe, and the cost of cementing it, needed to be recovered. To be economically viable, the washover operation cannot cost more than the cost of replacing the hole and the equipment lost in it.

EXAMPLE

In another job, 47 joints of 3½-in. drill pipe cemented in a 7-in. liner at approximately 14,000 ft., in a 36° hole was washed over. This job was completed successfully, but only 5 joints (approximately 150 ft.) of washpipe could be run at a time. When 6 joints were run, the string would stick, which added time and expense to the project in milling and fishing the washpipe. Another problem arose when the washpipe's rotation and reciprocation was stopped so that a wash-out tool could be run. The washpipe became stuck because of cement cuttings settling out.

When considering a washover operation, probability percentages must be applied to the cost formulas. Success is expected, but not all jobs are successful. What is the probability of success? Probability percentages should be determined from specific experiences in a number of similar jobs. Records from infield drilling and workover programs within the same field will indicate the problems experienced and their frequency. Only from experience in similar conditions can reliable probability factors be developed.

Type	Size															
TSWP	7-1/4	184.15	23.19	34.47	6.624	168.3	0.313	7.95	7.5	190.5	6.499	165.1	6-3/8	161.96	6-1/2	165.1
TSWP	7-3/8	187.3	28.04	41.68	6.63	168.3	0.38	9.5	-	0	6.5	165	6-3/8	161.96	6-1/2	165.1
TSWP	7-5/8	193.7	25.56	37.99	6.97	177	0.33	8.3	-	0	6.84	174	6-3/4	171.5	6-7/8	174.6
TSWP			29.04	43.16	6.88	174.6	0.38	9.5	-		6.75	172	6-5/8	168.3	6-3/4	171.5
TSWP			33.04	49.11	6.77	171.8	0.43	10.9			6.64	169	6-1/2	165.1	6-5/8	168.3
TSWP			29.70	44.14	6.843◆	173.8	0.38	9.5	8.01	204	6.75	17.2	6-9/16	166.7	6-11/16	169.9
TSWP			30.54	45.39	7.25	184.2	0.38	9.5	-	0	7.13	181	7	177.8	7-1/8	181
X-LINE	8	203.2	31.04	46.13	7.38	187.3	0.38	9.5	-	0	7.25	184	7-1/8	181	7-1/4	184.2
TSWP	8-1/8	206.4	35.92	53.39	7.25	184.2	0.44	11.1	-		4.13	181	7	177.8	7-1/8	181
TSWP			38.42	57.1	7.19	182.5	0.47	11.9			7.06	179	6-15/16	176.2	7-1/16	179.4
TSWP			33.95	50.46	7.58	192.5	0.4	10.1	-	0	7.45	189	7-1/4	184.2	7-3/8	187.3
TSWP			37.09	55.13	7.5	190.5	0.44	11.1			7.38	187				
TSWP	8-3/8	212.7	31.1	46.22	7.92	201.2	0.35	8.9	-	0	7.8	198	7-9/16	192.1	7-11/16	195.3
TSWP			36	53.51	7.813◆	198.5	0.4	10.2	9.12	232	7.7	196	7-1/2	190.5	7-5/8	193.7
TSS	8-5/8	219.1	39.29	58.4	7.73	196.2	0.45	11.4	-	0	7.6	193			7-11/16	195.3
TSWP	9	228.6	38.92	57.85	8.150	207	0.43	10.8	-	0	7.99	203	7-7/8	200	8	203.2
X-LINE			38.94	57.88	8.84	224.4	0.4	10	-	0	8.68	220	8-1/2	215.9	8-3/8	212.7
TSWP	9-5/8	244.5	42.7	63.46	8.76	222.4	0.44	11	-	0	8.6	218	8-1/4	209.6	8-1/2	215.9
TSWP			43.5	64.65	8.67	220.1			10.1	257	8.53	217				
X-LINE			46.14	68.58	8.68	220.5	0.47	12	-	0	9.79	249	9-1/2	241.3	9-3/4	247.7
TSWP	10-3/4	273.1	44.22	65.72	9.95	252.7	0.4	10.2	-	0	9.69	246	9-3/8	238.1	9-5/8	244.5
TSWP			49.5	73.57	9.85	250.2	0.45	11.4			9.6	244	9-1/4	235	9-1/2	241.3
TSWP			54.21	80.57	9.76	247.9	0.5	12.6	-	0	10.7	272	10-1/8	257.2	10-5/8	269.9
TSWP	11-3/4	298.5	52.57	78.13	10.9	276.4	044	11	-		10.6	270	10	254	10-1/2	266.7
TSWP			58.81	87.41	10.8	273.6	0.49	12.4			11.8	301	11	279.4	11-1/2	292.1
TSWP	12-3/4	298.5	49.56	73.66	12	304.8	0.38	9.5	13.5	343	12.3	311	11-1/2	292.1	12	304.8
TSWP	13-3/8	339.7	66.11	98.26	12.4	315.3	0.48	12.2	13.75	349	14.8	377	14-1/4	362	14-3/4	374.6
TSWP	16	406.4	81.97	121.8	15	381.3	0.5	12.6	16.75	426						

All strengths maximum value - apply safety factor of two (2) to the joint tensile yield strength.
* = Recommended make-up torque = 25% of maximum make-up torque - does not apply to X-Line Connections.
■ = Ratio of the joint tensile yield strength to the pipe tensile yield strength.
◆ = The internal upset has been reduced.
● = N-80 Pipe.
◊ = J-55 Material.

FIGURE 10-1

Washpipe data specification guide.

(courtesy of Baker Oil Tools)

FIGURE 10-2

Washpipe rotary shoe.

(courtesy of Baker Oil Tools)

will damage it. To prevent this, smooth brass is sometimes applied to the shoe's outside diameter as a bushing that reduces friction and prevents damage. Tungsten carbide is applied to the bottom of the shoe, and if possible, to the ID. A small carbide shoulder inside the shoe improves the chances of retrieving some of or all the fish. This can save a trip with another tool to recover the washed-over fish.

Because washpipe is large, stiff, and smooth, length of the washpipe string is a critical factor in preventing sticking. Length should be determined after careful consideration of hole conditions. Note that deviated holes or accidentally crooked holes limit the safe length of a washpipe string. The following discussions of two actual jobs demonstrate the extremes in washpipe-string lengths.

EXAMPLE

In one job, drill pipe in a vertical well was stuck from a depth of 330 ft. (the bottom of the surface pipe) to 8,487 ft. (the depth of the bit). The cause of the sticking was a poor mud system. The mud had to be circulated out and replaced, conditioning the hole in the process. On the last washover, 1,218 ft. of washpipe was run. This is unusual, but under the circumstances, the decision was correct and the job was successfully completed.

In this situation, equipment costs, as well as the cost of drilling the hole, the cost of the surface pipe, and the cost of cementing it, needed to be recovered. To be economically viable, the washover operation cannot cost more than the cost of replacing the hole and the equipment lost in it.

EXAMPLE

In another job, 47 joints of 3½-in. drill pipe cemented in a 7-in. liner at approximately 14,000 ft., in a 36° hole was washed over. This job was completed successfully, but only 5 joints (approximately 150 ft.) of washpipe could be run at a time. When 6 joints were run, the string would stick, which added time and expense to the project in milling and fishing the washpipe. Another problem arose when the washpipe's rotation and reciprocation was stopped so that a wash-out tool could be run. The washpipe became stuck because of cement cuttings settling out.

When considering a washover operation, probability percentages must be applied to the cost formulas. Success is expected, but not all jobs are successful. What is the probability of success? Probability percentages should be determined from specific experiences in a number of similar jobs. Records from infield drilling and workover programs within the same field will indicate the problems experienced and their frequency. Only from experience in similar conditions can reliable probability factors be developed.

When the entire length of the fish cannot be covered in one washover, it is necessary to separate the freed section of the string from the section remaining in the hole. This can be done using several methods.

- After pulling the washpipe, an overshot can be run, left-hand torque applied, and the fish backed off with a string shot (as discussed in Chapter 7).
- Instead of a rotary shoe, an external (or outside) cutter can be run on the washpipe and the fish cut off above the lowest point of the free section.
- A washpipe spear may be run in the washpipe string during the washover. The spear can then be used to apply left-hand torque for the string-shot backoff.
- A backoff connector can be run with a washover backoff safety joint and engaged at the top of the fish when the washover is completed. Through this connector, left-hand torque can be applied and the string-shot backoff made.

DIAMOND-DRESSED WASHOVER SHOES

Periodically, you may be required to mill or wash over items that are dressed with a wear protection material. These items could include rock bits, hole openers, whipstock faces, or cutting grades of tungsten carbide as found on milling tools, stabilizers, or even diamond products. When this application appears, there are different options for washover/milling heads. One example would be to use a diamond-dressed burning shoe (Figure 10-3).

Rotary shoes manufactured with diamond-impregnated heads can compete well with tungsten carbide and other hard metals and outlast conventional burning shoes. Their benefits include that they will stay in the hole and not wear out (as tungsten carbide against tungsten carbide will), and

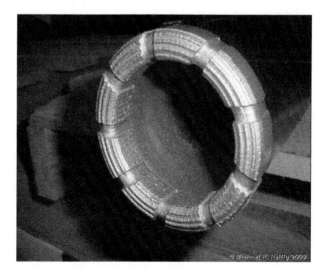

FIGURE 10-3

Diamond-dressed shoe.

(Courtesy of Micheal Riley)

reduced trips will be needed to replace worn shoes. As a result, one diamond-dressed shoe can do the work of many carbide shoes.

The disadvantages, however, are that they are expensive, slow, and less common than carbide shoes. They cannot be put into a grinding machine and modified to meet size requirements, nor can they be made in welding shops at local fishing tool companies. In addition, their slow performance often causes operators to apply too much weight due to impatience or inexperience, which can split or otherwise damage the shoe.

EXTERNAL CUTTERS

Outside or external cutters (Figure 7-6) usually have a slightly larger OD than washpipes, and it is dressed to catch the type of tool joints or couplings present on the fish. Three typical catcher types are used with this tool:

- Pipe with couplings requires a catcher assembly with spring fingers or flipper-dog catches that hook below the coupling. Once over the fish, as the washpipe is moved upward, the finger-catcher assembly near the top of the cutter engages the pipe under the coupling. In turn, this sleeve moves down the barrel, transmitting force through the spring to the sleeve, which feeds in the knives. As this assembly is rotated, the knives cut the pipe in two.

- Pipe with upset-joint couplings can be caught with flipper-dog or pawl catchers made with slip surfaces cut on the end that will engage the upsets. The pipe is then cut as discussed in the previous entry.
- Flush-joint pipe requires a hydraulically actuated catcher. Pump pressure against the sleeve restriction in the annular space actuates the knives.

Coil springs are used in practically all outside cutting tools to absorb heavy shocks that can break the knives.

WASHPIPE SPEARS

The washpipe spear, also called a *drill-collar spear* (Figure 10-4), is used mainly to prevent dropping a fish that is stuck off the bottom when washing over it. This versatile tool can be used to pick up a fish on the same trip as the washover and to back off a fish when it is partially washed over, saving a trip. When a fish is stuck off the bottom and it is washed over, it usually falls to the bottom and may corkscrew the pipe, damage the bit by knocking off cones or shanks, damage filter cake in the wellbore, or all three. Catching the fish before it falls saves considerable time and money.

The spear is made of two major assemblies: the mandrel-and-slip assembly and the control-cage assembly with friction blocks, restriction rings, and a latch. The spear is usually placed in the bottom joint of the washpipe string, but it can be run anywhere in the washpipe string. It is anchored in place by turning the bottom sub to the left, which moves the tapered slip cone under the slips, extending them and anchoring the spear in the washpipe. Below the spear, an unlatching joint or J-type safety joint is run.

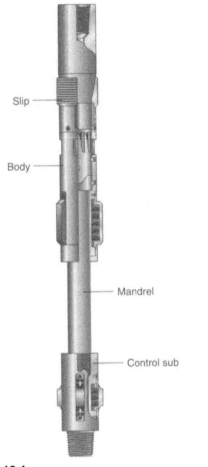

When the washpipe has been worked over the top of the fish and the spear engaged into the fish, the connection between the top of the fish and the bottom of the spear is fashioned by a rightward rotation. By continuing rotation with an upward pull, the slip cone is completely retracted, so the slips will not drag on the washpipe. The spear is now firmly connected to the fish and is not engaged in the washpipe except through the friction blocks on the control cage.

Normal washover pump pressure is applied to the surface of the restriction rings that hold the cage down on the mandrel. If the fish is cut loose and starts to fall, the friction blocks hold the cage firmly in the washpipe and the mandrel moves down. Without the cage holding the slips in a retracted position, a spring below the slips moves them up, engaging the washpipe and stopping any further downward movement of the mandrel and the fish.

Another type of spear has control dogs that sit in windows inside a control bushing. Once the fish is engaged and the connection tightened, a quarter-round of left-hand torque is applied, and the washpipe is moved down. At this point, the spear becomes part of the fish, and if the fish falls, slips on the spear will catch inside the washpipe, preventing the fish from reaching the bottom.

The washpipe spear can eliminate a stripping job when the fish is recovered. At the surface, when the fish is reached inside the washpipe, drill pipe may be picked up, made up handily in the top of the spear, and the spear manually latched off or disengaged. The spear with the fish is then lowered to the bottom of the washpipe and set in the last joint, after which the drill pipe is removed. The fish is now hanging out of the bottom of the washpipe, preventing a time-consuming stripping job.

When a fish is leaning over in a washed-out section of the wellbore and it is not possible to go over it with the usual washpipe assembly, a slightly bent joint of pipe can be run below the washpipe spear. With this pipe hanging below the washpipe, it becomes easier to stab the fish.

FIGURE 10-4

Washover washpipe spear.

(Courtesy of Baker Oil Tools)

Labels on figure: Slip, Body, Mandrel, Control sub

UNLATCHING JOINTS

The spear is always run with an unlatching or J-type safety joint immediately below it. The unlatching joint is held firmly in place with two light-metal straps, which prevent it from accidentally

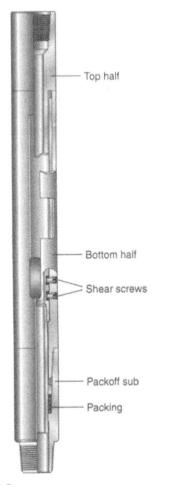

FIGURE 10-5

J-type safety joint.

(courtesy of Baker Oil Tools)

unlatching while going into the hole. After the spear and safety joint are made up in the fish, a straight pick-up pulls the straps apart and the safety joint is operational. It is usually dressed so that right-hand torque can be transmitted through it, a straight pull exerted on it, and unlatching accomplished with a slight left-hand torque while picking up.

The J-type safety joint (Figure 10-5) is a two-piece unit that is secured when lugs on the top half engage with a J-slot on the bottom half. Special shear screws keep the tool securely fastened until the fish is engaged. Once the fish is engaged, a tension load of 10,000 lbs. per shear screw is applied to the string. When the screws have been sheared and tension released, the string is elevated while applying a slight left-hand torque. The lugs then shift into the release slot, and the two halves of the safety joint separate. To reengage the safety joint, the string is lowered while rotating slowly to the right until the lugs on the top half engage the slots in the lower half. Rotation is then stopped, right-hand torque is maintained, and the string is lowered until the joint is fully engaged. The joint remains engaged until the releasing procedure is repeated.

BACKOFF CONNECTORS

When washing over and retrieving long strings of pipe resting on the bottom, a backoff connector may be used to reduce the number of trips with the workstring. In a washover and backoff operation, it is possible to wash over a stuck string, screw into the fish, and perform a backoff on the same trip. This assembly is essentially a J-type safety joint made up to a washpipe backoff safety joint (Figure 10-6). This can be made up to the top of a long string of washpipe or just run with a shoe and extension. The bottom of the J-type safety joint is subbed to the correct connection of the fish.

When the tool has screwed into the fish and it is established that the fish cannot be pulled, the J-type safety joint can be released to allow the washpipe to reciprocate and rotate. After a string shot has been run to just above the safety joint, the fish can be reengaged and the backoff completed. If the washpipe will not turn freely to the left, the anti-friction ring of the washover safety joint will allow this joint to unscrew. The washover safety joint's long, threaded area makes it possible to back off all tool-joint

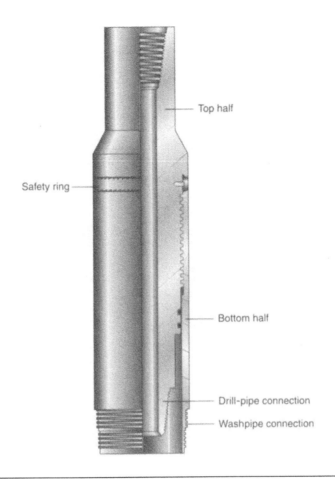

FIGURE 10-6

Washpipe backoff safety joint.

(courtesy of Baker Oil Tools)

connections and most tubing connections without completely unscrewing it, which in turn keeps the washpipe from unscrewing.

HYDRAULIC CLEAN-OUT TOOLS

During recovery of stuck drill pipe or tubing, the inside of the pipe occasionally becomes bridged over. This prevents the lowering of string shots, free-point tools, and electric wireline-cutting tools to the desired depth. Spudding with a wireline and an assembly on the bottom to cut through the bridge is sometimes successful if the bridge is not too long or compacted. When the bridge cannot be removed by spudding, the next step is usually a hydraulic clean-out tool.

The clean-out tool consists of a jet-type shoe, lengths of clean-out tubing (usually 1⅜-in. OD pipe), a top sub (which may include a stop ring and entry circulating ports), and a connection to the sinker bars and rope socket.

After a backoff or cut is made in the free pipe, the pipe is pulled and a circulating sub of the correct size and thread is installed on the bottom of the pipe string. The string is then run back into the hole and screwed into the fish. If the pipe is cut instead of backed off, an overshot without a packoff is used to tie back to the fish.

Up to 300 ft. of small tubing can be run into the hole, and when the bridged area is reached, the pumps are started. Jetting action washes away the plugging material as the clean-out tool is lowered on wireline. When this has been completed, the clean-out tool can be pulled from the well and normal pipe recovery operations restarted.

LOOSE-JUNK FISHING

The first step in fishing loose junk is to identify what it is. This may be readily determined if something has been left in the hole on a trip or has been dropped into the hole accidentally. If the type and configuration of the junk is not known, you should consider making a lead-impression block. If the fish can be identified, visualize how it may be retrieved by placing another part or tool that is exactly the same as the fish in a casing nipple of an appropriate size to simulate the hole. Then the proposed fishing tools may be tried at the surface, and inappropriate ones can be ruled out. It is much cheaper to try a tool on the surface than to make a trip and retrieve nothing. The usual tools used to retrieve loose junk are magnets, various types of junk baskets, hydrostatic bailers, and tools that might be devised for the particular circumstance.

MAGNETS

Fishing magnets are either permanent magnets, fitted into a body with circulating ports, or electromagnets, which are run on a conductor line. There are also inline magnets or down-hole magnets. Magnets will pick up only ferrous metal. Other methods should be used to recover brass, aluminum, carbide, and stainless steel.

PERMANENT MAGNETS

Permanent magnets have circulating ports around the outer edge so that the fill and cuttings can be washed away and contact can be made with the fish. Typically, the magnetic core is fitted with a brass sleeve between it and the outer body so that the magnetic field is contained and there is no drag on the pipe or casing. Permanent magnets allow circulation to wash away the fill so that the junk is exposed. Ordinarily, by rotation, the operator can detect when contact is made with the fish. The operator should then thoroughly circulate the hole, shut the pump off, and retrieve the fish. When pulling the workstring, it should not be rotated because there is a chance of slinging the fish off.

Permanent magnets are usually furnished with a cut-lip guide, a mill-tooth guide (which is the most popular), or a flush guide (Figure 11-1), which resembles a thread protector. A mill or cut-lip guide that extends below the magnet is extremely helpful in retaining the fish and preventing it from being dragged inside the casing.

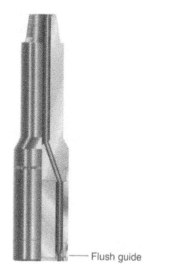
— Flush guide

FIGURE 11-1

Junk magnet with flush guide.

(courtesy of National Oilwell)

ELECTROMAGNETS

Electromagnets are run on a conductor line and are charged only when the bottom of the well is reached. They have the advantage of quick trips in and out of the hole, plus the added lifting power of an electromagnet. However, if the fish is covered with fill or debris, it cannot be reached because there is no way to circulate this tool down.

DOWN-HOLE OR INLINE MAGNETS

Down-hole or inline magnets (Figure 11-2) are added to a bottom-hole assembly to help remove ferrous-metal debris. They have a box-by-pin connection and resemble a drill-pipe pup joint. Between the connections are very strong magnets that attract metal chips from the wellbore fluid. The down-hole magnet can also be used to clean metal cuttings and other ferrous-metal debris out of casing collars, horizontal wellbores, and sub-sea blowout preventer (BOP) stacks. The inline magnets are used to assist with cutting- and milling-debris removal in situations with either low annular velocity or well fluids that have poor carrying capacity. If the down-hole or inline magnet is used in a milling or drilling operation, run it higher in the string to prevent extreme weight from being applied to the magnet. Should the magnets be used for recovering perforating-gun debris, they are run below the perforating guns.

JUNK BASKETS
CORE TYPE

For many years, a core-type junk basket was the standard tool for fishing-bit cones and similar junk from an open hole, and it is still frequently used. It consists of a top sub or bushing, a bowl, a shoe, and typically two sets of finger-type catchers (Figure 11-3). It is made to circulate out the fill and cut a core in the formation. The two sets of catchers, one dressed with short fingers, help to break off the core and retrieve it. Any junk in the bottom of the hole is retrieved on top of the core.

CATCHER TYPE

In all catcher-type junk baskets, the catcher must rotate freely in the bowl or shoe. As the tool is run over junk, the catcher snags on the junk and remains stationary as the bowl and shoe rotate around it. If the catcher is fouled with trash, excessive paint, corrosion, or other foreign material, it will not rotate and the fingers will break off, leaving additional junk in the wellbore.

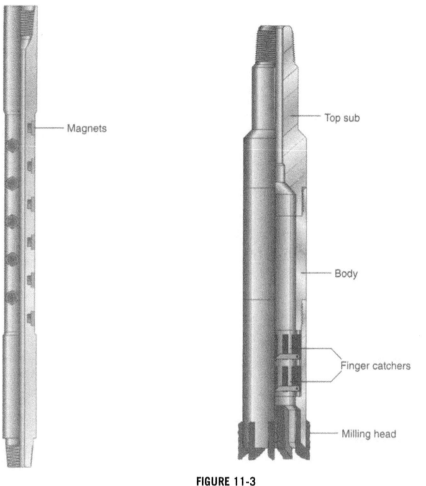

FIGURE 11-2

Down-hole or inline magnet.

(courtesy of Baker Oil Tools)

FIGURE 11-3

Core-type junk basket.

(courtesy of Baker Oil Tools)

REVERSE CIRCULATING TYPE

In many workover operations, fluid is "reverse-circulated" by pumping it down the annulus and return-ing it through the workstring. Because of higher fluid velocity through the workstring cross section, this circulates out larger and heavier particles than when pumping the "long way," or down the workstring with returning through the annulus. In many operations, the workover fluid need not be nearly as vis-cous if it can be reverse-circulated. In open holes, however, it is seldom possible to reverse-circulate due to problems caused by pumping into the formation. Nevertheless, the reversing action is extremely helpful in kicking junk into the barrel and catcher that might otherwise be held away from the catcher by the fluid flow. In recent years, two different designs of reverse-circulation junk baskets have been introduced.

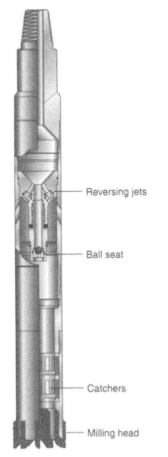

Reversing jets

Ball seat

Catchers

Milling head

FIGURE 11-4

Reverse-circulating junk basket.

(courtesy of Baker Oil Tools)

The first design incorporates jets, or *venturis,* that open when a ball is dropped down the workstring (Figure 11-4). The space behind a jet reduces the pressure. As the jet forces fluid from the bowl into the annulus, the interior of the bowl is at a lower pressure, which causes wellbore fluid to enter the basket through the catcher.

The other reverse-circulation design incorporates an inner barrel with the fluid flow between the outer and inner barrels. After washing the wellbore sufficiently to remove all fill, a ball is circulated down. When the ball seats, the fluid flow is diverted between the two barrels and flow is actuated through the upper ports into the annulus from the inner barrel. This creates reverse circulation in the immediate area of the junk basket.

Caution should be observed in dropping or circulating any ball or other restriction plug down the drill pipe or tubing. As mentioned in Chapter 4, it is critical to caliper, measure, and note the dimensions of all tools that are run down a hole. Some tools have restricted inside diameters (IDs) and will not allow balls or other items to pass. It should be standard practice to check this carefully before dropping anything down the workstring.

FRICTION SOCKET TYPE

Often, an off-the-shelf" or ready-made junk basket does not work because of the size and shape of the junk in the hole. In many cases, alternative solutions can be devised. The following are some examples:

- If the ID of the catcher is not large enough to accommodate the junk, a shoe or length of pipe may be used as the body of a shop-made junk basket. A series of holes may be bored or burned around the circumference of the material and steel cables brazed into place to form a catcher. It is not possible to rotate this tool on the junk, as the cables will be broken and torn out, but the tool can be pushed down over junk and the junk retained by the catcher.
- Cutting inverted "U" shapes into a piece of pipe and bending them in until they practically touch can also make a friction catch. This tool can be pushed down over a long, tubular piece of junk, and this technique is quite effective in cases where the junk's dimensions are unknown.

These two examples can address the two most frequent problems: junk that is too large to catch and junk with an unknown outside diameter (OD).

Several good "mousetrap" design tools have emerged in past years. One design, called a *kelo overshot* or *oversocket,* has tracks set on opposite sides and at an angle from the bottom to the top of the bowl (Figure 11-5). Various slips can be fitted to ride up and down on the beveled track. As the tool is lowered over the fish, the slip is pushed up until there is sufficient clearance for the fish to pass. Then the slip falls down and wedges the fish in the bowl. This tool cannot be released, but it is very effective for fishing sucker rods in casing or tubing that is so corroded that an ordinary overshot will not catch it.

ALLIGATOR GRAB

Although the alligator grab (Figure 11-6) is not used much in the industry today, it still has its place. For objects such as loose bits, bit cones, and odd-shaped items, using the alligator grab can be one option to retrieve them. There are two types of tools: one is mechanical and the other works off hydraulics and is mainly used in coiled tubing applications.

The mechanical tool is run into the hole in an open position based on the casing size. Once you reach the top of the object to be fished, you will slowly work over it and start rotating it to the right. At the point that you have achieved the

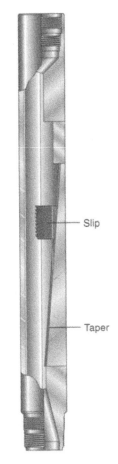

FIGURE 11-5

Kelo overshot.

(courtesy of Baker Oil Tools)

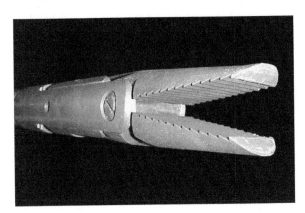

FIGURE 11-6

Alligator Grab.

(courtesy of Baker Oil Tools)

required amount of turns to fully close the tool or that you start seeing torque at the surface, pull it out of the hole slowly, without any rotation. Use caution as you come through the wellhead, blowout equipment, and the table.

Neither the mechanical tool nor the hydraulic tool is designed to use for heavy pulling or jarring, as you most likely will pull off your fish if it is required.

POOR BOY BASKETS

Poor boy baskets were among the first junk-basket designs used in the early days of the drilling industry. The poor boy basket is usually shop-made for a particular job from a short section of low-carbon steel pipe. Schedule-40 material is a good choice, but anything of a higher grade than J-55 will not work properly because the teeth will break off without bending.

The basket is made with teeth, cut with a cutting torch, and a curved leading edge. This edge is cut with a bevel. Note also that there is a gap between the teeth, and the teeth are about three-quarters of the diameter of the pipe from which they are made. This length of pipe is then threaded or welded to a top sub or bushing.

Running technique is critical to using this tool successfully. It is necessary to rotate and circulate the tool down over the junk without applying excessive weight. Because of the slots between the teeth, the tool will usually run roughly while the junk is at the level of the teeth. As the hole is made (by measurement) and the junk moves up into the smooth bowl of the basket, the tool will begin to run smoothly. When this has been accomplished, weight is applied as the tool is rotated, and the fingers will bend in and "orange-peel," retaining the junk inside the bowl. New teeth must be cut for each job.

BOOT BASKET

The boot basket is also called a *junk sub* or *boot sub* (Figure 11-7). It is designed to run in conjunction with and just above some other tool such as a bit, mill, magnet, or catcher-type junk basket. It will operate properly only when circulation is flowing the long way (i.e., down the workstring

FIGURE 11-7

Boot Basket.

(courtesy of Baker Oil Tools)

Hole Size or Pipe ID (in.)	Boot OD (in.)	Connection (API Reg)
4 ¼–4 ⅝	3 ¹¹/₁₆	2 ⅜
4 ⅝– 4 ⅞	4	2 ⅞
5 ⅛–5 ⅞	4 ½	3 ½
6–6 ⅜	5	3 ½
6 ½–7 ½	5 ½	3 ½
7 ½–8 ½	6 ⅝	4 ½
8 ⅝–9 ⅝	7	4 ½
9 ⅝–11 ⅜	8 ⅝	6 ⅝
11 ½–13	9 ⅝	6 ⅝
14 ¾–17 ½	12 ⅞	7 ⅝

FIGURE 11-8

Boot basket recommended sizes.

and up the annulus). Boot baskets can be manufactured with any connection, but most are made with a regular thread that will make up directly to a bit or mill without using a crossover sub. Washpipe boot baskets are made with a washpipe connection on the bottom and are run in place of the drive sub when run with washpipe or rotary shoes.

The boot on the basket is comparatively large for the hole or casing size (Figure 11-8), so high-velocity fluid return is accomplished through this portion of the string. As the fluid reaches the top of the boot, a much greater annulus area is present and the fluid pressure drops suddenly, creating turbulent flow just above the top of the boot. At this point, any heavy items such as steel cuttings, carbide inserts, bit teeth, or ball bearings circulating in this fluid will drop and fall into the boot. Boot baskets may be run in tandem to increase capacity, and some operators will also place another boot basket up the hole several joints to pick up junk that has been pumped higher than the lower basket.

Field welding should not be permitted on the mandrel of the boot basket. For example, if gussets are welded on the mandrel to reinforce the boot without stress relief, these welds may produce stress cracks. This can result in mandrel failure and an expensive fishing job.

HYDROSTATIC BAILERS

A hydrostatic bailer can be practical for cleaning out miscellaneous junk in the wellbore (Figure 11-9). Designs are made for running on pipe or wireline. All bailers work on the hydrostatic-head principle: They depend on the weight of the fluid in the hole to force the junk into the bailer and past the junk catchers. Many bailers can be surged repeatedly until the basket is full of junk or the hole is clean. They are particularly appropriate for cleaning out bit-cone parts, bearings, pipe slivers, bolts, nuts, gun debris, and other small pieces of nonmagnetic material.

FIGURE 11-9

Hydrostatic Bailer.

(courtesy of Baker Oil Tools)

JUNK SHOTS

Junk shots are large, jet-shaped explosive charges run on wireline or drill strings to break up objects left in the hole that are not recoverable by ordinary fishing methods. The charge breaks the junk into small pieces, which are typically recovered with magnets or junk baskets. Because the large charge creates a tremendous force, a cavern may be created, and sometimes debris is blown out into the sidewall of the hole. All the force of the explosion cannot be directed downward, even though the tools are designed so the maximum force is in this direction.

With any shaped charge, target distance is critical. For maximum efficiency, the charge should be circulated down to the fish if the junk shot is run on pipe. If the shot is run on an electric wireline, a bit run should be made to ensure that the charge gets completely down to the fish. A junk shot should never be run inside pipe, as the explosive force will destroy the casing or the pipe itself and probably increase the problem as a result.

MILLING OPERATIONS

MILLS DRESSED WITH TUNGSTEN CARBIDE INSERTS

Milling is the main method to remove objects that are stuck in a wellbore by friction, mud, cement, scale, or sand, or there is no tool that can grapple the object because of outside diameters (ODs) versus inside diameters (IDs) or odd shapes (i.e., bit cones). Successful milling operations require the appropriate milling tools, fluids, and techniques. The mills, or similar cutting tools, must be compatible with the fish materials and wellbore conditions. The circulated fluids should be capable of removing the milled material from the wellbore. Finally, the techniques employed should be appropriate to the anticipated conditions and the likely time allotted to achieve the objectives of the operation.

Since 1985, when Tri-State Oil Tools (Baker Hughes) invented carbide insert milling technology, there have been rapid changes in the design and usage of down-hole milling. This technology has changed the way that we think about and plan fishing and workover activities. With this and other advanced milling technology, it has been possible to mill strings of cemented tubular and sustain high milling rates of over 40 ft./h. The technology is based on using tungsten carbide inserts as the primary cutting element, with a composite of crushed tungsten carbide as a secondary cutting structure.

Over time, many other companies have developed the insert technology and advanced the original concept. They have refined many existing mill designs and conceived new mills and applications. This new breed of milling tools has established themselves in the inventories of modern fishing tool companies, but it has not replaced the technology that has been in use for decades before. There will always be a need for the crushed carbide tools that have been the solution to countless problems in the past.

Carbide insert mills give their optimal performance under very specific circumstances and fail miserably in others. Some basic criteria should be considered when determining whether to use a crushed carbide mill or a carbide insert mill:

- **Is the fish tubular?** Insert mills perform best milling tubulars.
- **Is the fish cemented?** Insert mills perform best on cemented tubulars.
- **Is the fish loose junk?** Insert mills do not perform well with loose junk such as bit cones, which should be left to the standard concave type junk mill.

Tungsten carbide insert mills generally do not perform well when the fish can move. The inserts tend to shatter and break when there is "chatter" (i.e., excessive vibration or jumping). The best applications for tungsten carbide insert mills are when the tubular is cemented in place. In many cases, a pilot mill (Figure 12-1) will be the best choice. In other cases, a flat-bottomed profile bladed insert mill would be preferable.

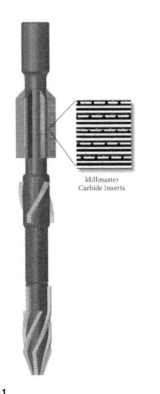

Millmaster
Carbide Inserts

FIGURE 12-1

Pilot mill with insert carbide.

(courtesy of Weatherford)

In order to achieve the best results, the bottom hole assemblies (BHAs) must be correctly configured, the operating parameters observed and adjusted for optimum performance, and hole cleaning best practices followed.

One of the contributing factors that led to the development of insert mill technology was the necessity to redrill depleted wells from offshore platforms to new bottom-hole targets using existing well slots. It was common to have wells with multiple strings of casing cemented together and extending below the intended kickoff point. Once the completion was pulled, the slot recovery procedure commenced.

Beginning with the remaining inner string of casing, a cement bond log should be used to determine the top of the cement. An internal pipe cutter can be used to cut the casing above the point where it is cemented, and then a spear and fishing assembly is used to retrieve the cut portion of casing. Alternatively, a one-trip system such as the cut-and-pull spear system can be used, potentially saving extra trips with cutter and spear assemblies.

Once casing can no longer be cut and pulled, milling operations can begin. Prior to selecting the correct mills to use, you should determine if the cemented casing is centered or eccentric in its outer string of casing. One way to determine the position is to run a lead impression block. Should the casing be centralized (Figure 12-2), an insert dressed string pilot mill can be used. The string mill gives a bit more versatility due to the lower rotary connection. The following bottom hole assembly should be considered:

One sample casing size is as follows: 7" x 32 lb-ft. inner casing and 9 ⅝" x 53.5 lb-ft. outer casing

Round-nose taper mill (5 ⅞" OD)
Integral blade stabilizer (IBS) (5 ⅞" OD)
String insert pilot mill (7 ¾"OD) - (⅛" Larger than coupling OD)
Integral blade stabilizer (IBS) - (7 ¾" OD)
Spiral drill collars
Drilling jars
Spiral drill collars
Hevi-Wate drill pipe (HWDP)

This assembly will guide and stabilize the pilot mill from the bottom and clear any swarf that may fall below the mill and cause a bride. The IBS above the pilot mill stabilizes the pilot mill from above and effectively prevents the assembly from chattering.

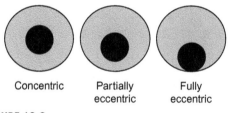

Concentric   Partially   Fully
               eccentric   eccentric

FIGURE 12-2

Casing concentricity.

Once the assembly is lowered into the well and the fish is tagged, the string should be elevated several feet and rotation and pumps started. The parameters such as weight, torque, SPM, pump pressure and RPM should be observed and recorded. Slack off the string until the mill starts to take weight and torque is observed. Allow the mill to slowly establish cutting and increase weight to optimize penetration rates.

Low weight on insert mills will wear them out prematurely, and the blades have a tendency to taper when too much weight is applied. Keep the weight at a minimum while maintaining the desired cutting rates.

Rotary torque is the key to high milling rates. A combination of weight and RPM should be found by varying parameters until a "sweet spot" is found, and the driller must pay close attention and make frequent adjustments to maintain the progress. To maintain optimal conditions, it may be necessary to pick up and work the string periodically to prevent "bird nesting" of cutting and rebedding the mill.

The tungsten carbide inserts are manufactured using very fine powder, which is formed into the desired shape in molds. They are heated and processed to produce a solid piece. The insert designs can incorporate custom features such as a chip breaker, which controls the size and shape of cuttings. A cutting angle can be built into the insert, negating the need to weld the blades onto the mill body at the desired negative rake, which enhances performance. Other features help to provide improved bonding and cushioning of the insert once it is fused to the mill blades by brazing with oxygen/acetylene techniques.

The amount of weight on the mill can be difficult to manage. Sometimes the desired penetration rates cannot be achieved with a low weight, and it must be increased until the casing starts to cut. This can vary by the weight and grade of casing, the hardness of the surrounding cement, casing accessories such as centralizers and scratchers, casing couplings, well deviation, centralized casing, and other conditions. Each milling job is unique; the mills are all handmade by individual artisans and are seldom identical or consistent in quality. Even the best tool dresser or carbide welder can have a bad day.

A steady creeping of the brake and not slacking off in large increments "chomping" on the fish should succeed at feeding the Mill. After a pilot mill run is completed, and before the next run is made, it can be beneficial to make a clean-out run with a flat-bottomed mill or a venturi-type reverse basket assembly to remove any remaining cuttings that were not circulated from the well, as well as any cement sheath left in the well. This is also a good time to condition the mud.

Cutting removal and management are also key elements to a good milling job. Do not mill any faster than the hole can be cleaned of cuttings. High penetration rates produce a large volume of cuttings, and the milling fluid and circulating system must be able to handle them. The following are recommended:

High yield point
High flow rate
High viscosity

Thixotropic properties
Gel water-based mud
Frequent high-viscosity sweeps
Periodic fresh drill water sweeps

While removing the cuttings (Figure 12-3), they should be checked for size, shape, weight, and volume. Returns should be inspected for cement and formation cuttings. If cuttings are long and stringy, it can mean that insufficient weight is being used. If they are thick and heavy, it could be that too much weight is being used. Ideal cuttings should be the length and thickness of a paper match, approximately 1 to 1 ¼" long, ¼" wide, and ⅛" thick. Mills can produce curly cuttings, which are difficult to manage because they tend to accumulate and pack off. The driller and tool operator (recommended) should keep a sharp eye on pump pressure. A slow, steady increase of pressure is an early indicator of packing off due to bird nesting. To prevent and cure bird nesting, the drill string is rotated and reciprocated and the spiral drill collars help to break up clumps of cuttings. If the cuttings cause the string to become stuck during the milling and cleaning operation, drilling jars can deliver both upward and downward force to free the stuck pipe.

Ditch magnets (Figure 12-4) should be placed in the mud cleaning system, which is located after the shaker screens, to catch very fine metal particles that have passed through the screens and prevent them from causing wear to the rig's mud pumps. It may be necessary to modify the system flow lines and remove sharp turns and bottlenecks that can cause cuttings to plug the lines. Trash skipping may be required for offshore operations in order to store cuttings.

If the inner casing string is cemented in an eccentric or offset, uncentralized position, using a pilot mill may damage both the outer casing and the mill. The force of the lower pilot inside the fully eccentric cemented casing could cause the pilot mill blades to cut the outer casing and bend or break the pilot mill blades while being pulled into the adjacent casing.

In this case, once identified, the milling assembly can be changed, and the lower mill replaced with a flat-bottomed bladed insert mill. Although milling progress will be considerably slower, the potential

FIGURE 12-3

Cuttings from pilot milling.

(courtesy of Michael R. Reilly)

FIGURE 12-4

Ditch magnets with cutting.

casing or mill damage will be avoided. Once milling has reached the shoe depth, the BHA can be changed, and a pilot mill can be run in the open-hole section, avoiding the possibility of sidetracking and benefiting from the higher milling rates afforded by the pilot mills.

During pilot milling operations when very high volumes of cuttings are produced, it is *not* recommended to run boot baskets or string magnets. This is because once they are full of cuttings, they will cause choke points for further cuttings.

Surface equipment is also a consideration in pilot milling. All 90° bends in flowlines should be eliminated, and a booster pump should be added at the bottom of the blowout preventer (BOP) stack. Flowlines should be opened to remove cuttings before they reach the shale shakers. Note that round cutters will produce 5 in.^{3} of cuttings for every cubic inch of steel milled. Even at 5 ft./h, a lot of cuttings have to be removed. See Chapter 17 for a discussion of surface equipment.

MILLS DRESSED WITH CRUSHED TUNGSTEN CARBIDE

Selecting the best mill for a specific task is not always straightforward. Many different mill styles are available. Some have multiple purposes, and others are suited for limited, specialty jobs. The following are guidelines and are intended as general rules.

Junk mills (Figure 12-5)
Flat-bottomed mill

These mills (Figure 12-6) are well suited for handling solid junk, cemented junk, squeeze tools, packers, and bridge plug milling. They can be spudded as required when junk rotates, and are generally solid, heavy-duty mills with stabilized mill head or body stabilizers. They have large circulating ports for powerful washing. Mill head segment separation can be large or small to permit or restrict large pieces, such as slip segments, to be lifted to the boot baskets. They are suitable for open and cased holes.

FIGURE 12-5

Crushed carbide junk mill.

(courtesy of Michael R. Reilly)

FIGURE 12-6

Flat-bottomed junk mill.

(courtesy of Michael R. Reilly)

CONCAVE BOTTOM MILL

The concave bottom mill (Figure 12-7) is excellent for milling loose junk, such as bit cones, reamer parts, scraper blades, tools, chain, dies, and float collar debris. Its concave form keeps junk centered for greater effectiveness. This heavy-duty mill is also suitable for dressing off fish and

FIGURE 12-7

Concave junk mill.

preparing for engagement with catch tools. It also has large ports to foster good circulation, and a well-stabilized body.

TOOTH-TYPE BLADED MILL

The pointed shape of the tooth-type bladed mill (Figure 12-8) enables a gouging, digging action that makes it suitable for handling cement, formation, sand, and general clean-out jobs.

FIGURE 12-8

Tooth-type bladed mill.

(courtesy of Michael R. Reilly)

TAPER MILL

The tapered mill (Figure 12-9) is versatile, with many cased hole and workover applications. It can be used as a pilot, guide, clean-out restrictions, perforations, run in combination with other string mills, boot baskets, and magnets. This type of mill is available in long or short styles.

STRING MILLS

The upper and lower connections on the string mill (Figure 12-10) permit it to be placed almost anywhere. It can be used for cleaning out wells, removing scales and burrs, opening restrictions, stabilization, polishing, and other purposes. This type of mill is ideal to run in conjunction with taper mills, and it can be used to ream keyseats and windows. In addition, string mills are excellent for cleaning "bird nests," and they have a smooth, machine-ground, full-gauge, center-bladed area to prevent damage to the casing ID.

ECONOMILLS

The economill (Figure 12-11) can feature either a flat or a concave bottom, with a round taper, bladed cement mill, or many other styles. It is available with crushed tungsten carbide or tungsten carbide inserts,

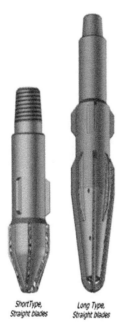

Short Type,
Straight blades   Long Type,
Straight blades

FIGURE 12-9

Taper mills, long and short.

(courtesy of Drillstar Industry)

FIGURE 12-10

String mill.

(courtesy of Michael R. Reilly)

FIGURE 12-11

Economill.

(courtesy of Michael R. Reilly)

and it is light, short and easily transported by air. The economill is commonly used as the internal mill of a three-piece skirted mill (described next).

SKIRTED MILL PREVENTS SIDETRACKING WHILE MILLING

The skirted mill (Figure 12-12) has three components: a drive sub, an internal mill, and a rotary shoe. The drive sub, which is also known as a *canfield bushing,* is a triple-threaded connection. The upper connection is fitted to the running string, a lower box connection secures the internal mill, and a lower pin connection screws into the rotary shoe.

The skirted mill lines up a damaged or distorted fish and slips over the fish, preventing it from jumping off. It then provides a sooth cutting action while dressing the top of the fish. Its demountable components allow changing out of worn parts rather than replacing a whole tool.

If you are running a shirted mill into cased holes, ensure that the tungsten carbide dressing on the OD of the shoe has been machineground to a smooth finish. The rotary shoe

FIGURE 12-12

Skirted mill assembly.

(courtesy of Schlumberger)

FIGURE 12-13

Watermelon mill.

(courtesy of Michael R. Reilly)

component can be replaced with a variety of guides, such as a cut-lip guide, oversized guide, or any other guide that meets the requirements of the operation. Before running the mill into the hole, ensure that the internal connection has been adjusted to the recommended torque, as they are sometimes shipped hand-tight only.

WATERMELON MILL

The watermelon mill (Figure 12-13) gets its name due to its shape. Often confused with the string mill, a watermelon mill has rough carbide dressing along the full arc of its OD. Its primary use is to elongate and ream the window mill during casing exit operations while departing from the whipstock. It is generally run above a window mill, and several may be run in tandem.

HOLLOW MILLS

A hollow mill (Figure 12-14) is constructed from sub stock with an API rotary shoulder connection at the top, and the ID of the bottom is machined

FIGURE 12-14

Hollow mill.

(courtesy of Michael R. Reilly)

out to the required size and depth. Most oilfield lathes can enlarge the ID to 15–18 in. deep. Afterward, the mill head is dressed with tungsten carbide in a fashion resembling a heavy-duty burning shoe.

Usually these mills are custom-made to meet special challenges. The mill can cut a fishing neck to enable engaging with a catch tool such as an overshot, thus "baiting" the fish. It can be used to cut cable that is fouling a cable head on a wireline tool. Often these mills will friction-grab the fish after dressing it off.

CONDUCTOR TAPER MILLS

The conductor taper mill (Figure 12-15) is used primarily when offshore driving and hammering of casing causes it to buckle or collapse and restrict the ID. An assembly consisting of increasingly larger mills can be run on a heavy drill collar string, beginning with a taper mill, followed by inter-mediate- and final-size string taper mills. Extremely heavy construction enables the tool to fulfill the tough task.

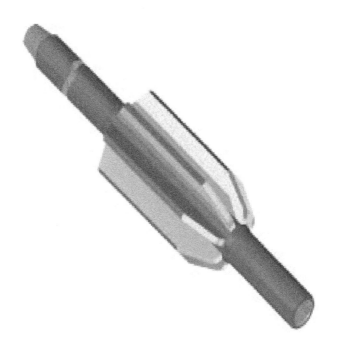

FIGURE 12-15

Conductor taper mill.

(courtesy of Schlumberger)

WIRELINE FISHING

One of the most challenging of all fishing jobs is the recovery of wireline and the tool or instruments run with it. The first consideration in wireline fishing is always whether the line has parted or it is still intact. There are also different procedures for electric (or conductor) lines and swab and sand lines.

If a conductor line has not parted, it is good practice to avoid pulling it out of the rope socket. If this occurs, contact with the tool or instrument will be lost, possibly permanently. If the instrument contains a radioactive source, the situation becomes even more critical.

To fish intact wireline, either the cable-guide method (better known as "cut and strip") or the side-door overshot method can be used. The cable-guide method should be chosen for all deep, open-hole situations, or when a radioactive instrument is stuck in the hole. This is the safest method and offers a high probability of success.

CABLE-GUIDE METHOD

A special set of tools is required for the cable-guide fishing method, and these are usually kept by the fishing-tool service company in a special box or container because they are not used for other purposes. The tools (Figure 13-1) consist of a cable clamp with a T-bar, rope sockets for each end of the line, one or more sinker bars, a special, quick-connector-type overshot for the line on the reel end, and a spear point for the well end. Also included are a slotted plate to set on top of the pipe, a sub with a recess or retainer to hold the rope socket, and an overshot to run on the pipe to catch the instrument or tool stuck in the wellbore.

A slight strain (approximately 2,000 lbs.) is taken on the line, and the cable hanger or T-bar is attached to the cable at the wellhead or rotary table. The cable is lowered so that the T-bar rests at the surface. The cable can then be cut so it reaches a convenient length above the floor. It is important to allow enough length because as in any deviated hole, the cable is pulled out from the wall, and more length is required to reach the surface than before the cable was stripped inside the pipe. Rope sockets are then made up on each end of the line, with the overshot on the upper end and the spear head on the lower end. As each stand of pipe is run, the cable spear-head rests on the C-plate to prevent the line from falling.

The first stand of pipe to be run (Figure 13-2) is made with an appropriate overshot on the bottom to catch the rope-socket fishneck or the body of the tool in the hole. It is always better to fish for the body of the tool rather than the rope-socket fishneck. Check to be sure that the guide or bottom does not have sharp edges that would cut the line if pipe weight were set down on it in a dogleg or on a ledge. A guide (called a *donut guide*) is usually run on the bottom of the overshot to prevent cutting the line.

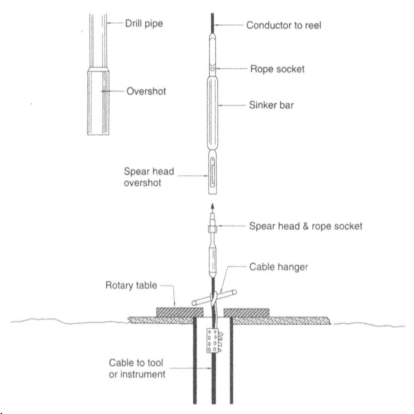

Drill pipe

Conductor to reel

Overshot

Rope socket

Sinker bar

Spear head
overshot

Spear head & rope socket

Cable hanger

Rotary table

Cable to tool
or instrument

FIGURE 13-1

Cable-guide fishing assembly.

The line to the reel is spooled up to the derrick worker, who then stabs the spear-head overshot and sinker bars in the pipe. With the pipe hanging in the derrick, the spear-head overshot is lowered through the pipe to the floor worker, who connects the spear head overshot with the spear point. The line is then picked up, and the stand of pipe can be run. This procedure is repeated until the overshot is about 30 to 90 ft. from engaging the wireline tool.

A sub that comes with the wireline kit has a recess inside, with a C-type insert. The spear-point rope socket may be stripped through this sub and the insert set in the recess. The spear point is then lowered onto the insert and disconnected from the spear-head overshot. Now the kelly or pump-in sub can be made up and circulation established to clean out the overshot. Caution should be taken to pick up or slack off on the line as little as possible while circulating. Picking up too much will pull the weak point out of the tool, and slacking off too much may cause the line to slack and ball up just below the overshot.

Once the overshot has been circulated cleanly, the sub with the insert can be removed, another joint or stand made up, and the fish engaged. Before moving forward, be sure that the fish is actually caught. The first check to make for this is to pick up the pipe. The line weight should start to slack off, as first will be seen on the wireline weight gauge, and then in actual slack in the line itself. The circulating sub

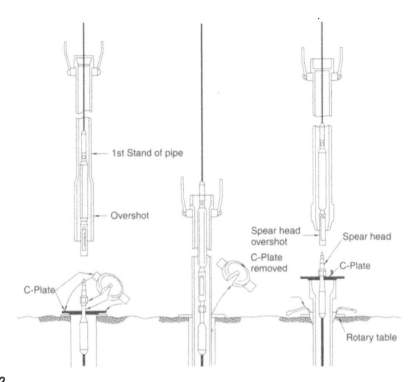

FIGURE 13-2

Cable-guide fishing method.

can also be picked back up and pressure applied to the fish in the overshot. This helps ensure that the fish is safely caught and will not fall out as it is coming out of the hole.

After the fish is securely caught, the T-bar clamp can again be placed on the line below the cut portion, the rope sockets removed, a square knot tied in the two pieces of line, and the line pulled out of the rope socket weak point with the elevator and T-bar. The line can be spooled up with the wireline reel, and then the pipe can be pulled out of the hole with the recovered instrument or tool.

As with all tools run in a well, wireline tools (including rope sockets, fishnecks, and instrument bodies) should be measured or calipered before running. If you are fishing these tools with the preceding method, the overshot above the grapple must be sufficiently open to swallow anything above the part being caught.

SIDE-DOOR OVERSHOT METHOD

The side-door overshot (Figure 13-3) is a special overshot with a gate or door in the side that can be removed to allow the line to be fed into the tool, after which the door is put back into position as part of the bowl. The overshot is run on drill pipe or tubing until the fishneck or body of the stuck tool is engaged.

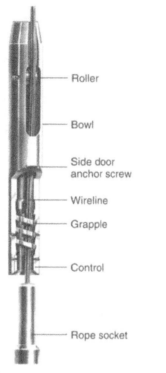

Roller

Bowl

Side door
anchor screw

Wireline

Grapple

Control

Rope socket

FIGURE 13-3

Side-door overshot.

(courtesy of National Oilwell)

The advantage of this recovery method is that the line does not need to be cut. It is also fast because no stripping is necessary. Be careful in setting the slips with the cable or line in the gap to prevent pinching or cutting. Because the line is outside the pipe, be careful not to rotate the pipe because doing this wraps the line around it.

Side-door overshots are not run in deep open holes because the line can become keyseated or even differentially stuck in the filter cake. All open holes are crooked enough for the line to drag along the side wall and cut a groove in the filter cake, causing excessive friction and sticking.

FISHING PARTED WIRELINE

Unlike a hemp rope or chain, wireline does not fall down the hole when it parts. It is stiff enough to stand surprisingly high. The larger the line, the stiffer it is, and the smaller the diameter of the pipe or hole, the shorter the distance that the line can fall. Because both factors can vary a great deal, it is not possible to formulate a rule of thumb, except to suspect that the top of the line will be higher than anticipated.

Rope spears or grabs are the most common tools used to fish parted wireline (Figures 13-4A, B). Each spear must be adapted to the size of the hole or pipe, and the barbs should be checked to be certain that the line will wedge tightly enough to pull it in two, if necessary. If the spear is run in casing or pipe, a stop ring must be run on top of the spear. The stop ring must be of sufficient size to prevent the line from going above it. This prevents catching the line low, which can ball up a long length and stick the fishing string. Always try to catch the line near the top. The more line that is pushed downhole, the more compacted the line becomes and the more difficult it is to penetrate and catch the fish with the barbs on the spear.

When it is not possible to make a good catch with a center spear, a two-prong grab rope spear (Figure 13-5) is usually used instead. This tool is used often for fishing slick line inside a tubing string. It allows the wireline to be caught from the outside instead of the inside.

If a spear is used to fish wireline in an open hole, no stop is run because the open hole is not always in gauge, and the line could pass the stop and get stacked up on top of it. This could cause the fishing string to become stuck and prevent it being pulled back into the casing shoe.

If the line becomes packed and impossible to penetrate with either the center or two-prong grab spear, it may be possible to screw a box tap on the ball of the wire. The box tap should be a full outside diameter (OD) body with a thin wall near the bottom. After the ball of line is fully engaged by the box tap, the line can be pulled in two, which provides a new fishing top.

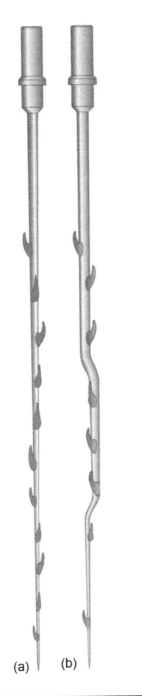

Always keep a tally of the amount of recovered line. This indicates the amount remaining in the wellbore. Because the line is usually unbraided, balled up, and no longer its original length, weight is often the only accurate method of estimating the amount of line recovered.

Because the line stands up in the hole, if the length left on the tool is short, even by as much as 100 ft., it still may be possible to fit an overshot to catch the tool and swallow the line. The overshot, extensions, and pipe above it should not have any small restrictions or square shoulders. The overshot should be slowly rotated as it is lowered. Short lengths of line can be caught with the tool, which is much easier than fishing the line itself.

CUTTING THE LINE

When a sand line or swab line is stuck in the hole, it is usually advisable to cut the line as low as possible so it can be recovered and the tool fished with a work-string of pipe. This is also advisable when a conductor line is run inside the pipe and the tool becomes stuck.

In the early days of cable tools, a rope knife was stripped over the stuck line and run on another line. Due to the lay of the two lines, the second line frequently became stuck, creating an even more serious situation. The explosive sand-line cutter has replaced this method.

EXPLOSIVE SAND-LINE CUTTER

The explosive sand-line cutter is dropped around the line. It is grooved, so it rides the line down to the top of the rope socket. A drop-weight bar slides down the line and hits the top of the cutter, which fires a small propellant charge. This drives a wedge that forces the knife to cut the line. Smaller-size cutters for tubing have fishnecks to allow retrieval with a pulling-tool run on measured line.

FIGURE 13-4

(A) Center-prong rope spear; (B) crankshaft rope spear.

(Both images courtesy of Baker Oil Tools)

(a) (b)

FIGURE 13-5

Two-prong-grab rope spear.

(courtesy of Baker Oil Tools)

In larger-size tubing, drill pipe, and casing, or in an open hole, a sleeve and sometimes guides are installed on the tool. The sleeve provides a seat for the cutting edge and allows a crimper to be installed, which clamps the line against the adapter sleeve. Then the cutter and drop weight can be recovered together on the end of the cut line.

Because the cutter is a free-falling tool, it is advisable to work the line to ensure that it falls as far as possible. This can be done by taking a strain on the line, releasing it, letting it fall 5–6 ft., and catching it with the brake. Shaking the line in this manner will work the cutter past some obstructions, but the cutter may stop on some splices or in mashed tubing. Wherever it stops, the cut will be made. There is no way of knowing where the cut has been made until the cut line is spooled up.

The explosive cutter is also available with an electronic timer programmed to fire the tool after a preset interval. It is used in wells where falling sand or solids tend to cover the cutter and prevent the drop weight from hitting the firing pin. It is also used in deviated wells where the drop weight's speed may be too slow to fire the cutter.

ELECTRIC SUBMERSIBLE PUMPS

When sucker-rod pumps and downhole hydraulic pumps can no longer lift sufficient fluid, electric submersible pumps are used. They consist of an electric motor, a pump, and usually some device for gas separation. To power the electric motor, an electric conduit must be run down. This cable is strapped to the production tubing with stainless steel, packing, and crate-type straps.

To contain a powerful electric motor and a pump, electric submersible-pump housings are relatively large in diameter and do not have much clearance between the pump and casing. Small deposits, such as sand and corrosion between the casing and pump, can stick the pump so it cannot be retrieved. Care should be taken not to pull the tubing in two because the cable, unlike other wireline, is flexible and quite heavy. If it parts, it falls and is easily packed down, so a center spear or other tool cannot penetrate it without breaking it into short lengths.

When pumps or the tubing become stuck, the tubing string should be free-pointed and the tubing chemically cut above the stuck point. Most installations include a check valve in the tubing string to prevent a backflow of fluid when the motor is turned off, which would cause damage to the motor. Free-point and cutting tools cannot be run below this check valve, so valves should be placed as low as possible in the production string.

Only a chemical cut should be made to part the tubing string, as this leaves a sharp cutting edge on the tubing, which is then used as a knife to cut the electric conduit. When the chemical cut has been made, the tubing will be parted, but the conduit (cable) will still be intact. The tubing string is then raised 18–24 in., providing a gap at the tubing cut, with the conduit pulled taut between the two sections of tubing.

A cutter with a bumper jar is run on a work-string of sucker rods or small tubing. The cutter is equipped with a spring-loaded arm that extends from the mandrel. When the cutter is measured into the depth of the gap, the arm extends, and as the tool is rotated to the right, the arm catches the cable and pulls it up against the mandrel of the cutter. By striking a series of blows with the bumper jar, the cable is severed with the sharp edge of the chemical cut. It is easy to determine when the cable is cut because torque in the work-string will be lost. When the cable is cut, the work-string with the cutter and then the production tubing with the electrical cable strapped to it can be pulled up. This will leave the pump and a short section of tubing and cable in the well. These should be fished with an overshot or spear and jarring assembly.

Depending on the design of the electric submersible pump, caution should be used in jarring upward. Some designs incorporate a flange on top that can be parted easily. Light blows should be used, both up and down, until some movement in the fish is detected. When this occurs, continued movement of the pump will work the fish free.

RETRIEVING STUCK PACKERS

As with any fishing job, it is useful to know as much about the equipment as possible when fishing stuck packers. If the make and model of the packer are known, data such as major dimensions, type, method of setting and releasing, and a picture or dimensional drawing can be obtained. Many times, it is beneficial to bring another packer of the same type to the well location. This provides immediate information if only a portion of the equipment is recovered and other parts remain in the well. Manufacturers' literature containing such information should be kept on file if possible because designs change and companies merge, are acquired, or go out of business. As a result, this literature may be the most accurate and readily available source of information about the stuck packer.

Packers generally fall into two categories: retrievable and permanent (the latter is shown in Figure 14-1). Retrievable packers include weight-set (with either J-Set or automatic bottom), tension-type, and rotation-set (with alternate weight and pickup on the setting string). Some retrievable packers have hydraulic hold-downs above the seals so that tubing weight does not have to be left on the pipe. Others are set hydraulically and must be released by rotation, shear pins, or rings. Some retrievable packers have special retrieving tools that sting into the packer and shift a collet inside that releases them.

Compared to retrievable packers, permanent packers are simple. They are configured with slips at the top to prevent them from moving up the wellbore after setting, a seal, and slips at the bottom to prevent them from moving down the hole. This configuration is sometimes modified by adding seal-bore extensions, mill-out extensions, and equipment such as blast joints and tubing below the packer.

RETRIEVABLE PACKERS

After the make and model of the stuck retrievable packer is determined, every reasonable effort should be made to release it before resorting to fishing. The tubing should be worked to ensure that it is completely free and that pipe friction is not adding to the problem. If the packer is released by rotation, torque must be worked down. To do this, a vertical mark is made on the pipe, and right-hand torque applied at the surface. While holding this torque, the pipe is reciprocated, which ensures that torque is distributed downward to act on the mandrel. Stretch should be measured and the depth of the highest stuck point estimated. At this point, running a free-point instrument should be considered.

If the slips are frozen, a string shot fired in the retrievable packer mandrel may reduce friction enough to free the packer. If formation conditions allow, pressure can be applied down the tubing and below the packer to create a lifting force. If solids have settled in the annulus, a hole can be punched in the tubing just above the packer mandrel and the wellbore circulated. If the packer is equipped with

FIGURE 14-1

Permanent packer.

(courtesy of Baker Oil Tools)

a hydraulic hold-down above the seal and slips, pressure can be applied in the annulus to help retract the hold-down buttons.

If the retrievable packer itself is stuck, a jarring string is usually effective. The tubing string is parted by either cutting or backing off the tubing just above the packer, usually a half-joint. The appropriate catch tool is then run with jars (see Chapter 8). When jarring, care should be taken to avoid jolting the mandrel out of the packer.

In some cases, the low-frequency, high-impact blows of a jarring tool can be a disadvantage. If a high-frequency, low-impact tool, known as a *downhole vibration tool* (Figure 14-2), is used instead, retaining forces such as friction and debris, which can cause the packer to be unreleaseable, can be overcome.

The downhole vibration tool has been highly effective in fishing retrievable packers. The fish is engaged, and a slight overpull is applied. Pump pressure is then started and the tool vibrates at a frequency of 12–18 Hz, causing approximately 1,100 impacts per minute. This tool eliminates the need for high-tensile work-strings or excessive overpulls to free the packer. This method also eliminates the need to pick up drill collars and the work-string does not need to be reciprocated up and down as with jars.

An alternative retrieval method is to wash over the packer and cut it out. This is suitable if part of the tubing string is stuck because of fill in the annulus. If only a short section of tubing is stuck and it is practical to wash over it in one trip, a dog-type overshot may be incorporated in the washover string. This overshot consists of a short section of washpipe (bushing) made up in the string, with an appropriate internal catcher to engage under the couplings of the tubing (similar to an external cutter). This could be a T-dog, flipper-dog, or rolling-dog overshot. Rotary shoes for this operation are typically of the tooth type, for digging out fill, mud, or cement. If the packer itself has to be cut, the shoe should have tungsten carbide on the inside and the bottom to mill away the packer's slips.

Some rules of thumb for fishing retrievable packers are as follows:

- Obtain or create a drawing of the packer.
- Know the location of the milling tool in relation to the packer parts.
- The outside diameter (OD) of the shoe should be just under the inside diameter (ID) of the drift of the casing size and weight.

- The shoe ID should be as large as possible, with a minimum of ¼-in. dressing.
- Have a relatively light weight (1–3,000 lbs.) and slow rotation (80–100 rpm) to minimize spinning.
- If it is necessary to spud on the slips and rings, do so lightly.
- If possible, burn over the top slips only.
- If rings begin to spin, use a light weight and high rotation (120 rpm or greater), and slow the pump down.
- Run bumper jars and at least two boot baskets.

PERMANENT PACKERS

Cutting over and retrieving permanent packers is a common and usually efficient operation. The tool string for this job consists of a carbide rotary shoe or mill, top sub or bushing, a length of small pipe used as a stinger, and a releasing spear or retriever (Figure 14-3).

The shoe or mill should have a smooth OD to keep from cutting the casing. It should be dressed with an ID small enough to cut away as much of the packer as possible without cutting into the OD of the mandrel. This will prevent any large pieces of junk from breaking up and getting remilled, which causes unnecessary damage to the shoe. The shoe should be long enough to cover the entire packer. If necessary, use an extension.

FIGURE 14-2

Downhole vibration tool.

(courtesy of Baker Oil Tools)

Retrieving tools come in several designs. Most come with some type of "J" mechanism on the grapple or inside the washpipe body, depending on whether the packer has a mill-out or seal-bore extension (Figure 14-4). If nothing is attached to the bottom of the packer, or if it has a mill-out extension, the retrieving tool will go through the bore of the packer and catch under its bottom.

Although it may be possible to pull the packer free after cutting only the top slips, it is advisable to mill through the bottom slips as well. This will help prevent the packer from hanging up while coming out of the hole.

If no provision has been made for a mill-out extension, a J-type mandrel must be run in the washpipe above the shoe and a spear fitted to it to engage the packer seal bore. The grapple on the spear may be pinned in the catching position with a small, brass shear pin. At no other time should spear grapples be pinned in the catching position.

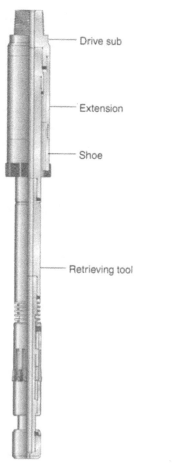

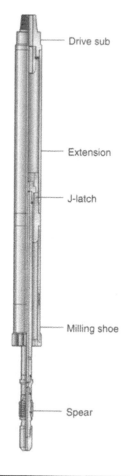

FIGURE 14-3

Packer-retrieving tool for a permanent packer.

(courtesy of Baker Oil Tools)

FIGURE 14-4

Packer-retrieving tool for a permanent packer without a mill-out or seal-bore extension.

(courtesy of Baker Oil Tools)

Newly developed retrieving spears will catch in the smooth bore of the packer and allow a packer mill to be run instead of a rotary shoe (Figure 14-5). This type of spear has a grapple that follows the mill down while milling, with the spear mandrel turning freely through the grapple. To release the spear, the mill is raised, a clutch located at the bottom of the spear mandrel is engaged, and the tool is rotated to remove the grapple from the packer. Milling packers with mills is faster than using shoes. This also creates smaller pieces of debris and allows for stronger connections, especially for packers set in smaller casings.

Some rules of thumb for fishing permanent packers are as follows:

- Obtain or create a drawing of the packer.
- Know the location of the milling tool in relation to the packer parts.

- Shoe or mill OD should be just under the drift ID of the casing size and weight.
- The shoe ID should be $\frac{1}{16}$ -in. larger than the OD of the packer mandrel body.
- Have a relatively light weight (1–3,000 lbs.) and slow rotation (80–100 rpm).
- Spud lightly on slips one or two times, and only if necessary.
- Slow down pumps if necessary to help burning over the packing element.
- If rings begin to spin, use a light weight, high rotation, and slow pumping.
- Run bumper jars and two or more boot baskets.

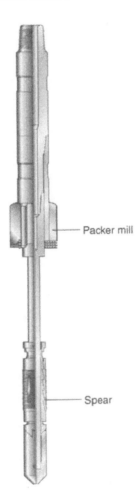

Packer mill

Spear

FIGURE 14-5

Packer-retrieving tool with packer mill for smooth-bore packers.

(courtesy of Baker Oil Tools)

FISHING IN CAVITIES

15

When drill-pipe parts in a washed-out section of the wellbore, the fish will not be centered in the hole. A straight-overshot tool string may bypass the top of the fish, touch the pipe, and take weight below the top. If this occurs, rotation slows, and the cut-lip guide will build up a slight torque and then "jump off." It may be impossible to engage the top of the fish with the tool string, but certain tools and techniques can be used to help latch onto the fish.

One of the simplest and most frequently used tools in this situation is a bent joint. A joint of pipe that is slightly bent just above the pin end and run just above the overshot will hang at an angle. By rotating it near the top of the fish, it may be possible to engage it. This setup is usually the first choice because it is simple and readily available on location. Care should be taken to bend the joint of pipe so that the opening of the cut-lip guide will be forced out toward the fish.

Some operators will run a jet sub, or kick sub, in place of the bent joint. This sub has a breakable plate just inside the pin end and a hole on the side. Pump pressure is exerted through this opening and against the wall of the hole, which kicks the tools to the far side of the wellbore. This is advisable only on limited occasions because the jet washes the sidewall, which removes the filter cake and erodes the hole.

Some subs are cut so the two ends are at a slight angle to each other. These are referred to as *bent subs, crooked subs, offset subs, angle subs,* and *degree subs.* They can be used instead of a bent joint.

If the bent joint alone does not allow the overshot to catch the top of the fish, a wall-hook guide can be substituted for the cut-lip guide on the bottom of the overshot. This guide is designed to catch the pipe below the top. Torque should be built up and held. Then, by slowly picking up the work-string, the fish is worked into the opening of the wall-hook guide and fed into the overshot bowl. Care should always be used in running a wall-hook guide because excessive weight or torque or running into a ledge can break the hook off.

If the wall-hook guide fails to catch the fish, a knuckle joint can be added to the string just above the overshot fitted with the wall-hook guide (Figure 15-1). Like a hinge, the knuckle joint moves in only one plane. The entire string—wall-hook guide, overshot, and knuckle joint—is put together and checked to ensure that the wall hook moves out with the opening facing forward when the string is rotated to the right. Shims are provided to adjust the wall-hook opening until it is in the proper plane. The knuckle joint swings free as it is run in the hole. Pump pressure against a restriction plug causes the overshot to kick out at a 7.5° angle. With the pump pressure holding the assembly out at an angle, the string is rotated to engage the fish. The restriction plug may be placed in the knuckle joint before running, or it may be pumped down the pipe to its seat. It is possible to get a very large sweep with the knuckle joint by adding extensions or pup joints between the overshot and the knuckle joint.

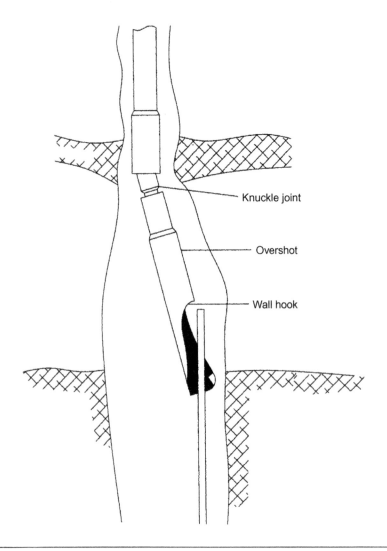

FIGURE 15-1

Fishing with a wall hook and knuckle joint in a cavity.

A knuckle joint (Figure 15-2) is inherently weak because it is a hinge. It will not withstand much jarring. Care should be taken when running it into the hole to avoid breaking it on a ledge. If the fish is engaged with this setup and it cannot be pulled, the restriction plug can be fished out with a small overshot run on wireline. This provides full opening of the tools. A free-point and string shot or cutter can then be run and the pipe parted below the tools, in a section of hole that is more in gauge. This freed portion of the fish can be removed and the overshot run back in without the knuckle joint so that the remaining fish can be jarred loose.

FIGURE 15-2

Knuckle joint.

It may be necessary to locate the top of the fish using a wireline induction log. Frequently, when fishing in an open hole that has washed-out cavities, the top of the fish will be some distance from the center of the wellbore. Because the pipe setting may corkscrew in compression, the top of the string may be difficult to find and identify.

SIDETRACKING METHODS

In recent years, new technology, such as three-dimensional (3D) seismic, has stimulated renewed exploration and development of mature oil fields. The ability to locate and produce previously undiscovered hydrocarbons in mature fields has led to increased drilling activity using sidetracking methods. In mature fields, plugging and sidetracking an existing wellbore is often more cost effective than drilling a new well. Offshore, the limited number of unused slots on a platform makes sidetracking the only feasible way to redevelop a field in many cases. Developments in tools and methods have also made sidetracking an economic alternative to conventional fishing jobs, and operators are choosing sidetracking over fishing more frequently than before.

When considering any type of reentry for sidetracking (Figure 16-1), be it whipstock, section mill, pilot mill, or cut-and-pull for slot recovery, it is important to communicate with the fishing-tool company, mud company, directional-drilling company, and wireline company for proper prejob planning. During well planning, communication between the fluid supplier and milling company is essential for an efficient milling operation that leaves a wellbore free of metal cuttings. Milling-fluid requirements vary widely between window-cutting, section-milling, and pilot-milling applications. Compared to window cutting, for example, pilot and section milling produce larger volumes of cuttings for removal by the mud system, so surface equipment must have larger flow lines, with a minimum of bends and irregularities, and should include centrifuges for pH control. Higher yield points, greater viscosity, and higher annular velocities are also required during section and pilot milling, creating the need for high-volume pumps. Effective preplanning for milling operations will ensure efficient removal of cuttings while keeping the volumetric ratio of cuttings to mud at an absolute minimum.

WHIPSTOCK SYSTEMS AND ANCHORS

Whipstocks are used to sidetrack a wellbore when reentering an existing wellbore as a cost-effective alternative to fishing a drilling bottom-hole assembly (BHA) that has become stuck, to address severe formation or drilling problems, to change the direction of a wellbore, or to drill multiple exits (multilateral wells) from a single wellbore. Whipstocks are available in permanent and retrievable versions.

HOLE CONDITIONS

To set a whipstock and sidetrack through casing, a good cement bond is critical. A cement-bond log should be run at the point of the proposed sidetrack. If the cement bond is poor, the casing could move during the milling operation, with unacceptable results. If the bond is inadequate, then the casing should

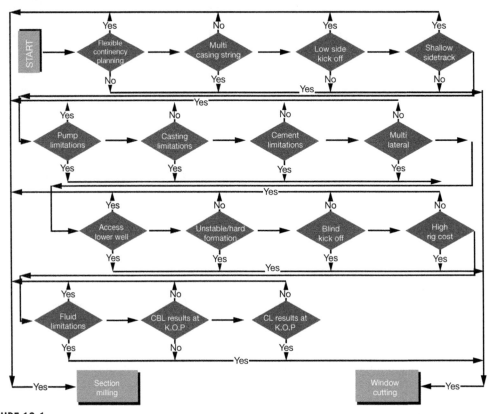

FIGURE 16-1

Identifying the appropriate sidetrack option.

(courtesy of Baker Oil Tools)

be perforated and cement squeezed around it. An alternate depth can be chosen where an adequate bond is found if doing this is practical and fits overall drilling objectives. It is also recommended that an anchor device be set 4 ft. above the casing collar to ensure that there is no attempt to mill through the collar. The casing collar can be found with a casing-collar locator run on wireline.

WHIPSTOCK-FACE ORIENTATION

Infrequently, an operator will have no preference as to the direction of a sidetrack, but for most side-track operations, proper whipstock-face orientation is a must. At the kickoff (K.O.P) point, if the well-bore has more than 2.5° of inclination, the whipstock should be oriented relative to the high side of the hole. Normal orientation of the whipstock face is recommended only within 60° to the left or right of the high side. Orienting the face outside this window increases the risk of window failure.

A whipstock assembly can be oriented using two methods: measurement while drilling (MWD) and a bypass sub. The MWD method can be used only in wellbores with an inclination of more than 5°. Also, a wireline surveying device, such as a surface-readout gyroscope (Figure 16-2) or steering tool, and a universal bottom-hole orienting (UBHO) sub can be used.

FIGURE 16-2

Surface-readout gyroscope.

(courtesy of Baker Hughes INTEQ)

CLEANLINESS OF CASING WALL

In some wellbores, especially those with high-temperature mud systems that have a tendency to settle out or cause corrosion, or those with older wells, the casing wall may not be clean. It is normally necessary to run a bit and scraper to below the packer/bottom-trip anchor setting depth to scrape the wall clean. When it is not clear that a casing scraper will clean the casing wall thoroughly, it may be necessary to use two or three full-gauge string mills to clean the casing at the kickoff point and allow proper packer/bottom-trip anchor slip setting. In cases where a whipstock packer will be set on wireline, run a wireline-gauge ring and junk basket to below packer-setting depth to ensure that the casing is full gauge.

FORMATION

The type of formation at the kickoff point can be a factor in the selection of milling equipment. Although most current milling systems can mill the window and drill the pilot hole in one trip, it is important to understand the formation at the kickoff point. When the window mill leaves the whipstock face, it leaves the milling mode, enters the drilling mode, and acts like a drill bit. In some formations, the mill does not drill the formation very efficiently, which requires another trip so that the pilot hole can be finished with a drill bit instead of the window mill. To prevent this extra trip, it is best to avoid very difficult formations whenever possible. Formations such as chert, some types of limestone, depleted sands, and granite can cause problems that result in an extra trip.

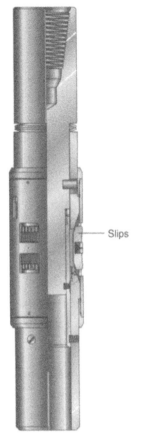

Slips

FIGURE 16-3

Bottom-set anchor.

(courtesy of Baker Oil Tools)

BOTTOM-SET ANCHOR

A bottom-set anchor (Figure 16-3) is used to anchor a whipstock in place in the wellbore. It is an alternative to the packer/anchor assembly. The bottom-set anchor attaches to the lower end of the whipstock with a drill-pipe connection. It can be used with a one-trip or two-trip milling system. To activate the setting sequence, only a bottom restriction is required. This can be a permanent or retrievable bridge plug, top of cement, liner top, production packer, etc. When using a bottom-set anchor, either the UHBO or MWD orientation method can be used.

TWO-TRIP MILLING SYSTEMS

The two-trip milling, or window-cutting, system is an efficient way to exit casing and provide a window suitable for running drilling BHAs, liners, and completion equipment. It is one of the most reliable systems in use, with thousands of successful field runs. The window is normally achieved in two round trips with drill pipe. The first trip lands the whipstock and makes a starting cut. The second trip completes the window and drills a pilot hole for the drilling BHA. When run with a whipstock packer, and if whipstock-face orientation is necessary, one MWD run or two electric-line runs are required to obtain the correct orientation.

The whipstock is run while pinned to the starting mill with a shear pin (Figure 16-4). The mill is made with a stinger, which serves two purposes. The stinger holds the whipstock, and it also guides the starting mill by keeping it inside the casing. This causes it to cut a long window instead of merely cutting a hole and going outside. Most shear pins are made to shear with approximately 20,000–45,000 lbs. of weight after the whipstock has been set. Once the pin has been sheared, rotation and circulation can begin, and the first phase of cutting the window is accomplished. The starting cut is usually around 12–24 in., depending on the design and size of casing to be cut. Mills used for this purpose are usually made with crushed carbide, insert carbide disks, or a combination of the two. Most operators do not run any drill collars with the starting mill because it should follow the taper of the whipstock.

The window in the casing is completed with a window mill, string mill, watermelon mill, or a combination of these (Figure 16-5). Cutting material is dressed on both the bottom and the sides of the mill. These mills are usually designed to ride alongside the face of the whipstock. An additional hole should be cut in the formation with this assembly so that the new hole is guided away from the old hole. The rule of thumb on any whipstock job is that after pulling the window, string, and watermelon mills

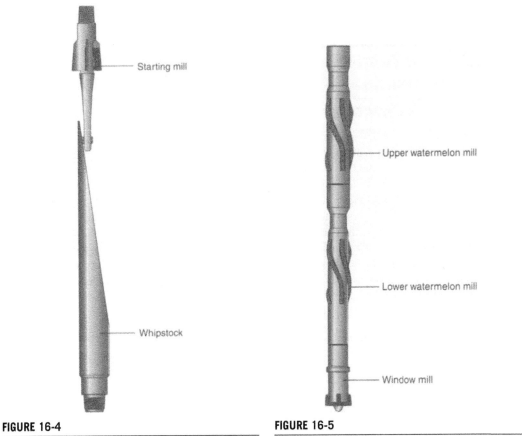

FIGURE 16-4

Starting mill pinned to the whipstock.

(courtesy of Baker Oil Tools)

FIGURE 16-5

Window-milling mills.

(courtesy of Baker Oil Tools)

from the hole, the outside diameter (OD) of the mills should be measured. If it is within the recommended gauge OD, drilling operations can begin. If the mills are under the recommended gauge, an additional milling run should be made to continue to open the window in the casing.

HYDRAULIC-SET WHIPSTOCK ANCHOR

The hydraulic, one-trip window-cutting system is used to exit casing efficiently and provide a window. The complete window is normally accomplished in one round trip with drill pipe. If a hole angle of 5° or greater exists at the kickoff point, an MWD tool can be used to orient the whipstock face. A bypass valve is required between the MWD tool and the whipstock. The valve allows flow to the annulus for MWD readings. Once the whipstock is properly oriented, the bypass valve can be closed hydraulically. As the valve stops flow to the annulus, it simultaneously provides fluid passage to the window mill. The hydraulic whipstock includes a steel tube inserted into the window mill that provides a means to apply hydraulic pressure to a packer. The packer can now be set hydraulically, without manipulating the drill string. The packer includes a rupture disc to prevent accidental release. Once the packer is set, the window mill is released from the whipstock by shearing down and then pulling off of the steel tube. Then the milling process can be started.

ONE-TRIP MILLING SYSTEM

The one-trip milling, or window-cutting, system (Figure 16-6) is used to exit casing efficiently and provide a window suitable for running drilling BHAs, liners, and completion equipment. The complete window is normally accomplished in one round trip with drill pipe. In one trip, the starting cut is made, the window milled, and a pilot hole drilled for the subsequent drilling BHA.

When run with a whipstock packer, and if whipstock-face orientation is necessary, two electric-line runs are normally required. The first is to set the packer, and the second is to determine the direction of the orientation key located inside the packer. When run with a bottom-trip anchor, only one electric-line trip is normally required. This is typically an electric-line gyroscope tool run through the drill pipe and into a UBHO sub located above the milling BHA, with its internal key previously lined up with the whipstock face. The desired whipstock-face orientation in this case is obtained by drill-pipe manipulation.

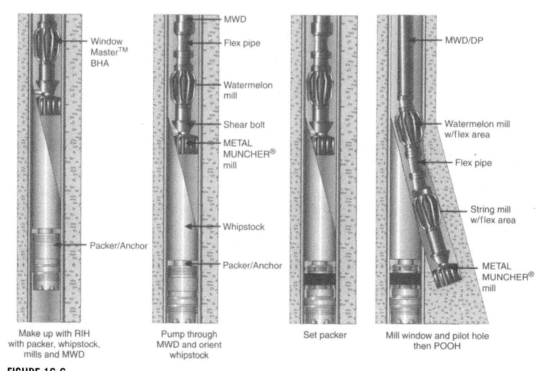

Make up with RIH with packer, whipstock, mills and MWD — Pump through MWD and orient whipstock — Set packer — Mill window and pilot hole then POOH

FIGURE 16-6

One-trip window-milling system.

(courtesy of Baker Oil Tools; WindowMaster™ trademarked by Baker Oil Tools; METAL MUNCHER® patented by Baker Oil Tools)

When using the bottom-trip anchor, if an MWD tool is available and the hole angle is 5° or greater at the kickoff point, the use of electric line can be completely eliminated. Instead, run the MWD tool in place of the UBHO, with its tool face previously lined up with the whipstock face. Pump flow through the MWD tool then will provide the whipstock face direction at the surface. In this case, only one drill-pipe trip is necessary to run in, orient, and anchor the tool; mill the window; and drill the pilot hole, making it the most desirable method when applicable.

Unlike conventional and two-trip systems, the one-trip milling system can start, mill, and ream the window without requiring a change in BHAs. A preliminary "starting-mill" run is not required. The result is a complete, full-gauge window in one milling trip. The one-trip milling system can be used with most currently available anchoring systems.

COILED TUBING SIDETRACKING

Coiled tubing drilling can be used when it is not economically feasible to utilize the main platform rig or mobilize a drilling rig. Advances in seismic and reservoir modeling are also helping to target smaller pockets of oil, further enhancing the economics of coiled tubing use. One of the operations essential for a successful and cost-effective coiled tubing sidetrack is the casing exit, discussed next.

CASING EXITS

Coiled tubing casing exits do not require the well to be killed and can be conducted in underbalanced conditions. The operation uses the existing completion to produce the well, including safety valves. In addition, smaller volumes of drilling fluids are required, and little formation damage is caused.

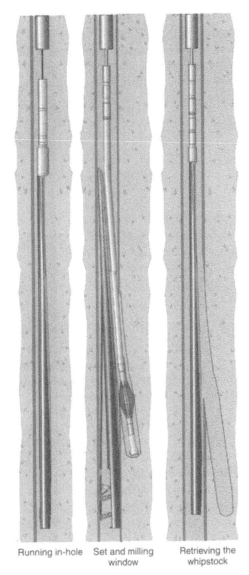

Running in-hole   Set and milling window   Retrieving the whipstock

FIGURE 16-7

Thru-tubing whipstock system.

(courtesy of Baker Oil Tools)

Currently, two coiled tubing casing-exit methods exist, which depend on the completion configuration; through unrestricted monobore completions and thru-tubing casing exits.

The monobore whipstock, also referred to as an *in-tubing whipstock,* can be set inside the tubing or in a monobore liner and the sidetrack performed by milling a window in the liner or through the tubing and outer casing. This method uses a traditional combination of packer and oriented-whipstock ramp. By contrast, the thru-tubing whipstock (Figure 16-7) is run through the tubing, and the whipstock is set in the larger casing or liner below it. It has no packer, allowing it to be set in a single trip.

The sidetrack is usually performed in a single milling run using a tungsten carbide or diamond window-milling assembly.

Both systems have been used successfully in a range of applications, with the whipstock deployed on either coiled tubing or electric line.

SECTION MILLING

17

Section milling and pilot milling (discussed in Chapter 12) are similar in that both operations mill away large sections of casing or pipe, but they are not the same. Section milling creates an exit point in a string of casing that is equal to the bit size of the casing being milled (Figure 17-1). Pilot milling completely removes an entire casing string from the top to the point where milling stops. The resulting hole is larger than the outside diameter (OD) of the casing being milled.

SECTION MILLS

Down-hole section milling of casing is generally done for the following reasons:

- To mill away the perforated zone in an oil string, which permits underreaming and gravel packing or completion in the open hole.
- To mill away a section of casing for a sidetracking operation. A new hole may be started in any direction because a full 360° opening is provided.
- To mill away a loose joint of surface pipe.
- To blank off a storage zone in a reservoir by removing a section above and below it and squeeze-cementing the storage zone.
- To cut pipe downhole for any purpose.

In sidetracking operations, 75–100 ft. of casing is ordinarily milled away. This is sufficient for easy exit from the pipe. Casing collars or couplings can be located by extending the blades of the section mill with light pump pressure across the section mill as it is lowered into the wellbore. Less weight on the indicator means that the blades are in the recess of the coupling. A rule of thumb is to start the outward cut about 5 ft. above a coupling. If the casing has little or no cement on the outside, this will help prevent the casing joint from backing off.

Section-mill cutting blades are operated by pump pressure (Figure 17-2). Blades are available in a variety of shapes such as round, rectangular, or triangular buttons (Figure 17-3). All blades are made of tungsten carbide. Current designs can cut a 100–150-ft. window in one trip.

A piston and cylinder in the section-mill body responds to pump pressure to force the blades against the pipe. The pipe is cut through, the blades extend through the gap, weight is applied, and the casing is milled. Drill collars are always run above the section mill to stabilize it and give the operator weight control.

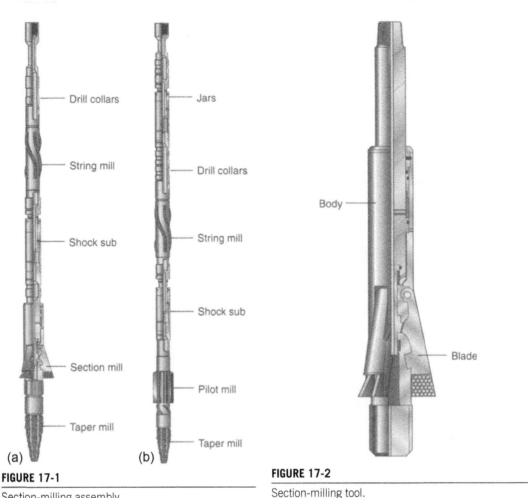

FIGURE 17-1

Section-milling assembly.

(courtesy of Baker Oil Tools)

FIGURE 17-2

Section-milling tool.

(courtesy of Baker Oil Tools)

Circulating a specialty milling-type fluid or mud formulated for the purpose is necessary to remove cuttings from the well. New cutting-structure designs produce faster cutting rates, so new cutting fluids have been developed to remove the cuttings more efficiently. One such specialty fluid is a mixed-metal hydroxide. If a water-based mud is used, it should have a funnel viscosity of 90–100 CP and a yield of 50–60. If the viscosity is too high, the mud will channel, especially in high-angle wells.

Surface equipment is also a consideration in section and pilot milling. All 90° bends in flowlines should be eliminated, and a booster pump should be added at the bottom of the blowout preventer (BOP) stack. Flowlines should be opened to remove cuttings before they reach the shale shakers (Figure 17-5). Note that round cutters will produce 5 in.^{3} of cuttings for every cubic inch of steel milled. Even at 5 ft./h, a lot of cuttings have to be removed (with the setup shown in Figure 17-4).

FIGURE 17-3

Tungsten carbide buttons.

(courtesy of Baker Oil Tools; METAL MUNCHER patented by Baker Oil Tools)

FIGURE 17-4

Special surface-equipment setup for section milling or pilot milling.

(courtesy of Baker Oil Tools)

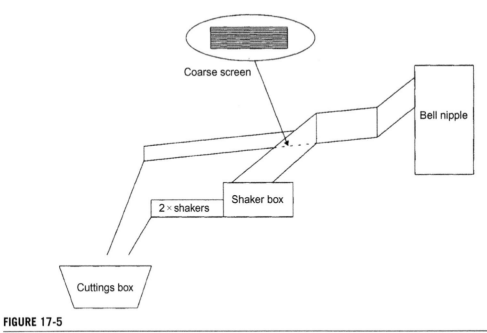

Coarse screen

Bell nipple

Shaker box

2 × shakers

Cuttings box

FIGURE 17-5

Example of flow lines and surface equipment.

REPAIR OF CASING FAILURES AND LEAKS

18

Leaks in casing have many causes. Among them are bursting or collapse from excessive pressure, thread leaks from improper makeup, corrosion holes, erosion from tubing leaks, and perforations that are longer than needed or desired.

The type of leak and its severity dictate the method of repair. The exact location of the damage must be determined, which is usually done by pressuring between a bridge plug and a retrievable packer or pressuring between a retrievable packer and the blowout preventers (BOPs). The packer is moved until the hole or leak can be accurately located, and then the best repair method is chosen.

Squeeze cementing is probably the most common method of sealing leaks in casing. Cement is pumped out through the leaks and allowed to set. The repair is then tested. Sometimes it is necessary to stage the cement job and leave some cement in the casing under pressure until it sets. In these cases, the cement plug is drilled out before the repair is tested.

A liner can be set to blank off a section of leaking casing. It can be set all the way to the bottom of the hole and hung in the casing above the leak (the same as in an open hole). Liner hangers can incorporate a packer to seal at the top of the liner between the liner and the casing, while other methods, such as squeeze cementing, depend entirely on cement. Liners reduce the inside diameter (ID) of the casing, which restricts operations and equipment. This factor can rule out the use of a liner in many cases.

If the leak is high in the wellbore and it is not practical or economical to set a liner all the way to the bottom of the well, a "scab" liner can be set across a short section of the casing that includes the leak. With this method, the liner is set on the slips of a liner hanger with a packer. The top is equipped with a setting sleeve and a liner packer with hold-down slips. This assembly will pack off the top and bottom of the casing section and isolate the leak. A disadvantage to this method is that a smaller-diameter section of casing is left in the well, with larger-diameter casing below it.

If the cost is justified, leaking or badly corroded casing can be removed by cutting the string below the damaged section, removing it, and running new casing with a patch to tie it back to the string left in the well. First, the lowest leak point is located or a casing-inspection log is run to determine the lowest depth of deteriorated casing. Then the casing is cut with a mechanical or hydraulic internal casing cutter (Figure 18-1) run on a work-string of tubing or drill pipe. The inside cutter is run to a depth below the damaged casing and rotated to the right, which releases the slips. Then, as slight weight is applied to the string, the knives are fed out on tapered blocks and rotation cuts the casing. The cutter is then pulled, and the cut casing is removed from the wellbore with a spear.

Casing patches normally have a slightly larger outside diameter (OD) than a standard coupling for the same casing size. To ensure proper sealing and fit, a casing dress-off tool should be run (Figure 18-2).

FIGURE 18-1

Mechanical internal cutter.

(courtesy of Baker Oil Tools)

FIGURE 18-2

Casing dress-off tool.

(courtesy of Baker Oil Tools)

There are many variations of dress-off mills and shoes. Ideally, the one chosen should be the same OD and length as the patch being run; this ensures that the patch will fit in the wellbore where it is to be set.

There should be dressing or hard banding inside and at the bottom of the tool, which should be dressed to an ID slightly larger than the OD of the casing. This will clear away any cement or dried mud on the outside of the casing to be patched. Finally, a cutting structure should be located inside the top of the tool to slightly bevel the inside and outside of the casing to be patched. This allows the patch to pass over the casing easily without damaging the packing inside the patch. Once the dress-off tool is run and the casing is properly dressed off, the casing patch is run.

Casing patches (Figures 18-3A, B) are made in several styles. Neoprene and lead are the two primary seal materials. The neoprene type is more commonly referred to as a *rubber (or neoprene) seal*. The rubber seal is rated for withstanding higher pressure than lead, while the lead seal has a higher temperature rating and is considered more resistant to corrosion.

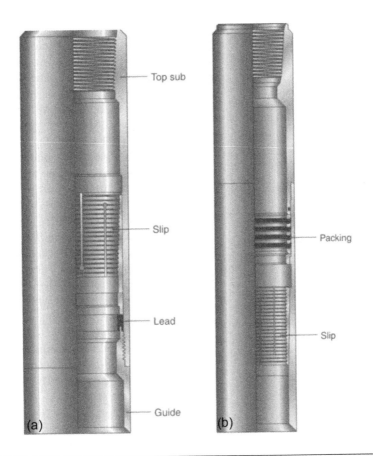

FIGURE 18-3

(A) Lead-seal casing patch; (B) rubber-seal casing patch.

(Both images courtesy of Baker Oil Tools.)

Casing patches are available for a variety of applications. One style, the cementing-type lead seal, permits the displacement of cement outside the casing and patch before it seals off (Figure 18-4). Either the lead- or rubber-seal patch is available with a long extension on top for salvaging pipe that has become stuck before landing on a subsea wellhead. This is called an *underwater casing patch.*

When using an underwater casing patch, the casing is cemented in place on the bottom to prevent it from falling down the hole if it comes free. A cut is made a few joints below the subsea wellhead, and this casing is recovered. The new casing top is then dressed off, and exact measurements are taken to determine the location of the casing top. These measurements are critical for spacing out the casing hanger, casing, and patch to be run back into the hole.

An underwater casing patch, made of either lead or rubber, is then attached to the bottom of the new casing string. The correct length of casing is run and the casing hanger and running tool are installed and run in the well. Approach the top of the casing stub with caution. After slacking off over the stub

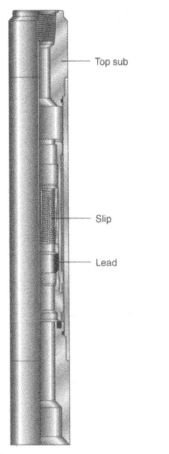

Top sub

Slip

Lead

FIGURE 18-4

Cementing-type, lead-seal casing patch.

(courtesy of Baker Oil Tools)

and seating the casing hanger in the wellhead, the hanger running tool should be retrieved. Next, a casing spear is run inside the casing and through the casing patch. The casing is engaged below the patch and pulled up to the required tension.

External casing patches are full opening and full strength, which are significant advantages. Once they have been engaged and the casing landed, they should be pressure-tested to ensure that a seal has been made.

Internal casing patches are designed to set inside the bad joint of casing to seal off perforations or other small leaks. These are used only if future well operations can be performed with a restricted ID in the casing string. There are several types of internal casing patches being sold today. The first is a patch consisting of three basic components: a top, soft-metal element with a rubber seal; a tubular extension; and a bottom, soft-metal element with a rubber seal. Various patch lengths can be obtained by adding tubular extensions. Only the top and bottom elements are swaged outward during the setting process. The tubing or casing extension is not swaged.

During the setting process, the pressure-setting tool exerts a pushing force on the top, tapered swage and a pulling force on the bottom, tapered, metal swage. Both swages are driven into the soft-metal sealing elements, which expand into metal-to-metal sealing contact with the casing bore.

Another option, the HOMCO internal steel liner casing patch, is a thin-walled steel liner that tightly molds to the inside of casing with the intent of permanently sealing any type of casing leak. The standard patch restricts the ID of the casing by 0.300 in. A heavier patch is also available in some sizes of casing with a 0.480 ID restriction. Depending on the area that is being covered by the HOMCO patch, you can achieve internal pressures up to 9,850 psi; however, external ratings are very limited in pressure integrity and can be as low as 400 psi, based on casing size and area that is being covered.

The standard HOMCO patch arrives at the location in corrugated form (which is star-shaped in cross section) with fiberglass cloth on the outside. When starting the patch in the hole, epoxy is applied to a fiberglass cloth. The fiberglass cloth, along with the epoxy, act as a gasket. Once in position, pulling a specially designed expander assembly through it with a hydraulic cylinder assembly expands the patch. Once expanded or formed against the casing ID, it is permanently held in place by radial compression.

Note: *Please contact your internal casing patch supplier to discuss all operational and internal and external pressure ratings prior to choosing which patch to run.*

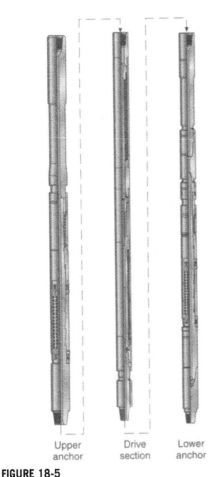

Upper   Drive   Lower
anchor   section   anchor

FIGURE 18-5

Casing backoff tool.

(courtesy of Baker Oil Tools)

If the casing is premium grade or thread, it may be desirable or necessary to back off the old casing and replace it. This method ensures that the string retains its original strength and pressure ratings.

Backing off the old casing is done by cutting it in the middle of the casing joint to be removed or cutting it just below the hanger to remove strain from the casing string. Once the cut has been made and the cut casing recovered, an internal spear is run on a work-string. The casing is engaged, and a wireline string shot is run and positioned across the connection to be backed off (see Chapter 7). A left-hand torque is applied from the surface, and the string shot is fired. The casing is then backed off completely and removed from the wellbore. A new casing string is run into the hole and the connection made to the old casing in the wellbore. The hanger is set, and the casing is tested.

Sometimes a string shot will swell or split the casing connection. To avoid this, a specialty tool called a *casing backoff tool* (Figure 18-5) can be used to unscrew the failed string at the desired connection. The casing backoff tool consists of a bottom anchor, drive section, and top anchor. The failed casing string is cut in the middle of the casing joint to be removed. The casing backoff tool is run on a work-string with a casing spear just above it. It is good practice to run 10,000 lbs. of drill-collar weight below the tool to aid in setting the bottom anchors.

The casing backoff tool is positioned with its bottom anchor below the connection to be unscrewed and its top anchor above the connection. Pump pressure is applied to set the bottom anchor and then the top anchor. Once both anchors are set, pump pressure is increased and the tool will stroke a half-turn to the left. Each time the tool strokes, the pressure is bled off and reapplied. This procedure is repeated until the casing has unscrewed, approximately three to five rounds.

The casing backoff tool is released by picking up on the work-string. After the tool is released, the work-string is lowered down the hole until the casing spear is engaged. Additional left-hand torque may then be applied to finish unscrewing the casing. The work-string is pulled from the hole along with the recovered piece of casing. New casing is run into the hole and made up to the old down-hole casing. The casing can then be tested and the casing hanger set.

The new casing string is usually run in with just a standard joint of casing on the bottom. In some cases, however, it may be necessary to use a tieback alignment bushing (Figure 18-6) on the bottom of the casing string to help screw the casing back together. The tieback alignment bushing will line up the

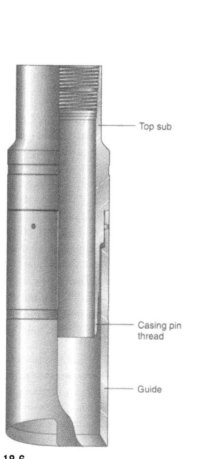

FIGURE 18-6

Tieback alignment bushing.

(courtesy of Baker Oil Tools)

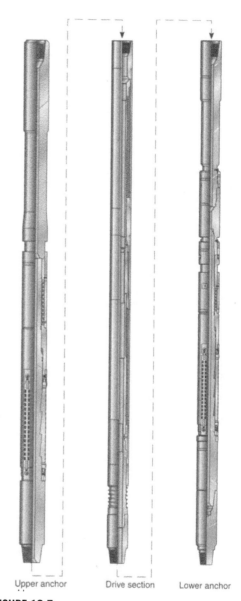

Upper anchor   Drive section   Lower anchor

FIGURE 18-7

Casing make-up tool.

(courtesy of Baker Oil Tools)

connections to prevent cross-threading the casing threads. The bushing is available with either a pin or a box connection.

When running a premium-threaded casing string that has critical requirements for makeup torque, another specialty tool called a *casing make-up tool* is used (Figure 18-7). This tool is similar in operation to the casing backoff tool, but it strokes to the right. It is positioned across the connection to be made up, and as pressure is applied, the casing connection can be set to the exact make-up torque recommendation. This is ideal for maintaining the integrity of premium-threaded casing.

Collapsed casing is another cause of casing leaks and failures. Collapsed casing can be difficult because it is not always easy to determine the severity and extent of the damage. Repair operations should be carefully planned and executed as follows. First, determine the length of the collapse from the allowable movement of the tubing and measurements correlated by free-point instruments. This determination is critical because the tubing must be cut both above and below the collapsed section.

When the electric wireline cuts are made, enough free pipe should be left above the collapse for an easy overshot catch. Enough should be cut below the collapse to prevent the tubing left in the wellbore from becoming an obstruction when swaging or rolling the casing. The cut section of tubing should be caught with an overshot and jarred out, typically with a string made up with a bumper jar, oil jar, drill collars, and intensifier.

After the cut section of tubing is removed from the wellbore, an impression block should be carefully run in to take an impression of the collapsed pipe. The impression block should be measured outside the hole, as the depth is always critical.

If the collapse runs up the hole from the most severely damaged point, the impression block will be depressed on one side, indicating that it was wedged into a taper (Figure 18-8). This type of collapse does not present any special problems, other than its severity. If the collapse begins at a coupling and extends down the wellbore, the impression block will be marked only on the bottom, at the point where it was pushed down on the end of the collapsed joint (Figure 18-9). This indicates that the collapse

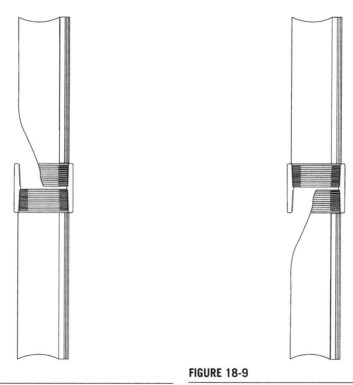

FIGURE 18-8

Pipe collapsed upward from coupling.

FIGURE 18-9

Pipe collapsed downward from coupling.

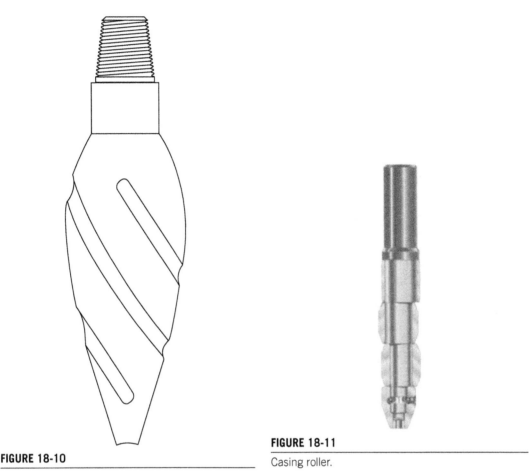

FIGURE 18-10

Casing swage mandrel.

FIGURE 18-11

Casing roller.

(courtesy of National Oilwell)

down hole is similar to a whipstock, and any tools run into the hole may have a tendency to go outside the pipe. Carbide mills should be avoided when repairing casing unless it is impossible to reform the pipe. Tapered mills tend to "walk," and in this situation, they will invariably follow the path of least resistance and go outside the pipe.

Casing swages or swage mandrels (Figure 18-10) are heavy, tapered cones that are driven down through the collapse and jarred back out. It is usually necessary to run a number of sizes in sequence because the pipe must be swaged out in small increments, sometimes as little as ¼ in. at a time. Most collapsed casing can be swaged out to approximately ⅛ in. below the drift diameter.

The first casing rollers (Figure 18-11) were adapted from swage mandrels by adding a series of rollers. Improvements have made them more reliable. Jars and drill collars should always run in either procedure because the swages and rollers can get hung and wedged in the collapsed section and must be jarred loose.

Once the casing is opened up sufficiently for normal operations, it must be reinforced in some manner before exposing it again to the external pressure that caused the collapse. This can be done by squeeze cementing or setting a liner through the repaired section.

FISHING IN HIGH-ANGLE DEVIATED AND HORIZONTAL WELLS

Successful fishing is much easier in a relatively straight well than in a highly deviated wellbore. However, it is still possible to perform a successful fishing job in a highly deviated wellbore if the proper approach is taken. Industry knowledge gained by personnel who have drilled many highly deviated wells, along with continuing developments in tools and techniques, have substantially reduced the number of fishing jobs required per well drilled and increased the success rate of those that are performed.

When a highly deviated well does require a fishing job, most of the tools used in straight-hole fishing can be successfully run. Even washpipe, with its specialized connections, can be run in highly deviated wells. Because the pipe is still large and not very flexible, sections must be short to pass through high-angle doglegs. Jars, overshots, magnets, and junk baskets can also be used.

There are considerations for job planning on a highly deviated well that are not present in fishing vertical holes. They include the following:

- When a high-angle hole has been drilled by rotating drill pipe, a trough usually forms on the low side of the hole that is smaller in diameter than the drilled portion of the hole. This is a factor when fishing with an overshot or similar tool, as the fish will lie in the trough or smaller section.
- Broken or twisted-off pipe can fall under a ledge of a dogleg.
- Hole drag in horizontal or highly deviated wells inhibits good jarring action.
- Adding weight to the string for milling operations can cause problems. Drill collars are similar to washpipe, in that they cannot flex around high-angle doglegs and bends. They have to be run higher up the hole, which reduces their effectiveness.
- It is difficult to get the required torque down and around deviations when attempting to back off pipe in a highly deviated well.

Just as drilling highly deviated wells has become a mature industry, so has fishing them. Applying the lessons learned by personnel on many previous jobs to your planning, as well as taking advantage of the latest advances in tools and techniques, will maximize your ability to overcome these challenges successfully.

PLUG-AND-ABANDON OPERATIONS

Plug-and-abandon operations are usually governed by the relevant regulatory agency of the country or region where the well is located. Regulations specify such things as the location and number of plugs to be set and the depth of the tubing and casing strings to be removed. Regulations may vary according to the well's location on land or its water depth offshore.

These operations can be either temporary or permanent. Not all the steps mentioned in this chapter may be needed in every well. In either procedure, the first step is killing the well and pulling the production string, if applicable. This can be accomplished by pulling all the tubing and packers from the wellbore or by pumping cement down the production string and into the producing zone. A bridge plug can then be set in the tubing string. Next, the tubing is cut at a depth that is predetermined either by the operating company or by government regulations, and another bridge plug is set in the casing on top of the cut tubing string.

At that point, all remaining casing strings and the conductor pipe are cut and pulled according to applicable regulations. On land jobs, cutting the casing strings is usually done by digging out around the casings at ground level and cutting them off with a torch.

TEMPORARY PLUG-AND-ABANDON

Offshore exploratory wells are often temporarily plugged and abandoned after drilling. The type of temporary plug-and-abandon process discussed in this section usually applies to wells drilled from a jackup-type rig. When a producing zone is discovered, the well is temporarily plugged until a platform can be put in place to accommodate production equipment. A platform is usually set over the first well, and more wells are then drilled from that platform.

On offshore exploratory wells, a combination of cement plugs and bridge plugs is set in the liner or main casing string. A surface plug is set just below the mudline-hanger system, which is designed to allow the unscrewing of the casing strings from a depth below the mudline back to the surface. When all the plugs have been set, the blowout preventer (BOP) stack is removed and the casing cut below the casing hanger. A spear is run, and the casing is engaged at the top. Most mudline hangers release with right-hand rotation, so it may be necessary to run an internal catch spear with vertical wickers (Figure 20-1) to aid in turning the pipe. Once the pipe is engaged, a slight overpull is taken. The pipe is then unscrewed at the mudline-suspension hanger (located about 150 ft. below the mudline) and the casing pulled from the wellbore. A corrosion cap is then run back into the well to protect the mudline-suspension hanger threads and prevent trash from falling into the wellbore.

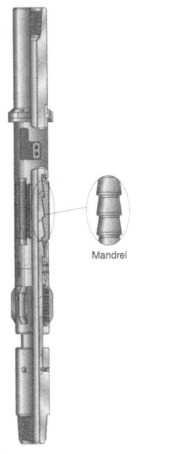

Mandrel

This process is continued for all casing strings other than the outermost casing or conductor pipe. The conductor pipe is then cut about 5 ft. above the mudline, with either an internal cutter or by divers. Some conductor casings have connections that can be unscrewed. Divers install bolts in these connections to allow the pipe to separate. Another corrosion cap is then installed over the entire well.

Once the platform is ready to be installed, the corrosion cap is pulled and the platform is set over the well. The conductor pipe is either screwed back together or a larger size is driven over the existing casing. The inner corrosion caps are then pulled, casings are run and screwed back into the mudline-suspension hanger, and the casing is hung off at the surface in the wellhead.

PERMANENT PLUG-AND-ABANDON

The first few steps of permanent plug-and-abandon operations are very similar to the temporary type. The well is killed and the production zone is squeezed or cemented. After the tubing string or strings are removed, all the casing strings are cut at a minimum of 15–25 ft. below the mudline or in accordance with applicable regulations. First, the innermost casing string is cut with an internal casing cutter. The casing is then removed and a bridge plug is set on top of the cut casing string. This process is repeated for all casing strings.

Usually, the casing strings are cemented together for well control or support. Therefore, they have to be removed at the same depth, which is typically done with a multiple-string casing cutter (Figure 20-2). This tool can cut multiple casing strings in one run. The cutter can be dressed with knives that are long enough to cut through the largest diameter casing string. A typical string might include 9⅝-in., 13⅜-in., 16-in., and 30-in. casings. In some cases, it may be necessary to run a shorter knife and then follow up with a longer knife on an additional run to complete the cut.

In deepwater operations off floating rigs or platforms, the tools become more specialized. Bridge plugs and cement plugs are set much the same way as a land or shallow-water plug-and-abandon job. Once these plugs are set (according to applicable regulations), the casing strings are cut. Because of rig movement, casing cutters must be stabilized and held in a stationary position during this operation. A marine swivel (Figure 20-3) is used to locate accurately and hold the multistring cutter in a stationary position. The swivel is placed in the subsea wellhead on top of the casing hangers. A slack-joint

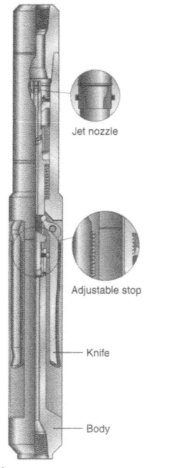

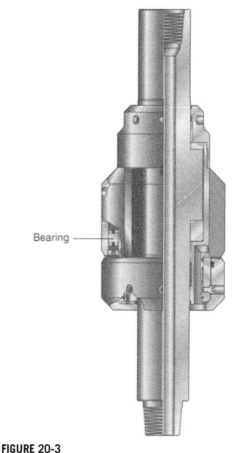

FIGURE 20-2

Multiple-string casing cutter.

(courtesy of Baker Oil Tools)

FIGURE 20-3

Marine swivel.

(courtesy of Baker Oil Tools)

or long-stroke bumper jar is then run directly above the swivel, with the cutter spaced out below the swivel. The slack joint or jar is then closed a half-stroke, which allows for rig movement but holds the swivel in a stationary position.

The swivel consists of bearings or bushings that will allow rotation of the inner body while the outer body remains stationary. In some cases, a seal puller can be adapted to the swivel, which allows pulling of the hanger seals on the same trip as cutting the casing.

When all intermediate casing strings have been removed, the conductor pipe can be cut and the subsea wellhead recovered. With cut-and-pull technology, you can cut and recover the subsea wellhead in one trip (Figure 20-4). Today, there are tools that cut with the drill string in compression or tension. Cutting and recovering the wellhead in tension provides for less drill-pipe fatigue and a positive

FIGURE 20-4

Subsea wellhead-retrieving spear.

(courtesy of Baker Oil Tools)

indication of the cut. Multiple-string casing cutters with sleeve-type stabilizers attached to the body also boost efficiency for quicker cutting.

Explosives can also be used to sever casing and casing strings for some plug-and-abandon operations. They can be run on pipe or wireline. Regulations concerning the use of explosives can be strict. Explosives work best if cement is between the casing strings to be severed. Jet-cutting and abrasive-cutting systems are also used for platform removal and plug-and-abandon operations. Jet cutting uses an abrasive agent such as sand or grit in water under high pressure. These tools can quickly and easily cut multiple strings of cemented pipe.

MISCELLANEOUS TOOLS

The miscellaneous or specialty fishing tools described in this chapter are common, but there is not much need for their use. When the need does arise, it is difficult or impossible to accomplish with other tools what these specialty tools are capable of doing. They should be kept in mind when planning a fishing job.

MOUSE TRAPS

A *mouse trap* is a catching tool with a moveable slip that allows a variable catch to be made. Ordinarily, this tool does not release, so it has limited applications. The advantage is that it will catch fish of variable or unknown size. Mouse traps are commonly used to catch sucker rods (Figure 21-1).

Larger versions of mouse-trap catching tools are typically used to catch corkscrewed rods in casing and tubular fish with inconsistent or nonstandard diameters, such as corroded or mashed pipe. One such large mouse-trap tool is the socket, also called an *oversocket* or *Kelo socket overshot,* which was originally designed for cable-tool use. It consists of a bowl, sized for the casing, and two tracks set at an angle, running from the top to the bottom of the bowl. Slips are fitted in the tracks and are free to slide up and down. The fish pushes the slips up the slanted track until the clearance is sufficient for it to pass the slips. When this occurs, the slips fall behind the fish and wedge it into the bowl. The fish can then be retrieved or pulled in two. A mouse trap–style overshot should never be run and latched onto a fish that is stronger than the overshot. If the fish is not free, it must be pulled in two.

REVERSING TOOLS

Reversing tools (Figure 21-2) are used to unscrew and recover sections of right-hand pipe or tools stuck in the casing. They are available in mechanical and hydraulic types. The hydraulic type generates 15,000–50,000 lb-ft. of torque, depending on tool size.

The reversing tool is used with a right-hand work-string. Planetary gears and an anchoring system convert the right-hand rotation on top of the tool to left-hand rotation below it. The gear ratio is about 1:1.6 and decreases as overpull is applied. At around 20,000 lbs. of overpull, the ratio is 1:1.

The reversing tool is for cased holes only. It should never be run into an open hole. If the fish is in an open hole, left-hand drill pipe must be run below the reversing tool. Reversing tools have restricted internal diameters, but the opening is usually large enough to accommodate a string shot if the fish cannot be unscrewed.

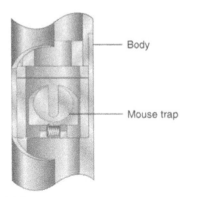

FIGURE 21-1

Mouse trap.

(courtesy of Baker Oil Tools)

HYDRAULIC PULLING TOOL

The hydraulic pulling tool (Figure 21-3) uses hydraulic pump pressure to pull objects from cased holes. It is designed to anchor in casing, exert a pulling force on the fish below it, and transmit this force to the casing rather than to surface equipment. The hydraulic pulling tool is highly effective for exerting unusual forces when using conventional workover rigs and small work-strings. The tool may be used for pulling liners, retrievable packers, seal assemblies, or pipe from a well, without strain on the work-string or derrick. The pulling tool is composed of three sections:

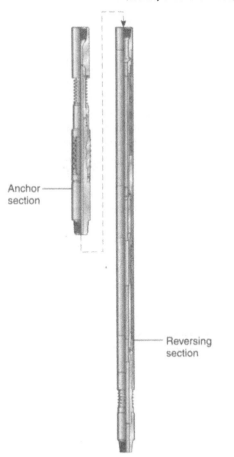

FIGURE 21-2

Reversing tool.

(courtesy of Baker Oil Tools)

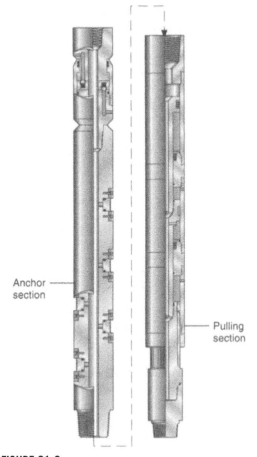

FIGURE 21-3

Hydraulic pulling tool.

(courtesy of Baker Oil Tools)

- The *relief valve section* is run on top of the tool and allows it to release the slips and the pull load whether or not differential pressure exists at the tool.
- The *anchor section* is run immediately below the relief valve and anchors the tool to the casing wall. It provides a large slip-engagement area to safely transmit heavy pulling strains with minimum risk to the casing.
- The *pulling section* is run below the anchor section, and a blanking sub is run below the pulling section so that hydraulic pressure may be applied. The pulling section produces forces ranging from 35 lbs. of pull per 1 psi (with a 4-in. tool) to 106 lbs per 1 psi (with a 7-in. tool).

The pulling tool can be used with mechanical fishing tools such as overshots, spears, or screw-in assemblies. It should always be anchored several joints above the casing shoe because there is an equal force pulling down on the casing when it is in use. Catching tools that are run with the pulling tool should be spaced apart with suitable drill collars or other heavy pipe. Bumper jars and safety joints should always be run below the pulling tool and directly on top of the catching tool. The bumper jar will allow the operator to bump down and stop the freezing of a grapple or a spear. The safety joint will allow the operator to unscrew the pipe to the left should the bumper jar fail to release from the fish.

The pulling tool has a 30-in. stroke. If the fish will stretch this amount or more before it reaches its breaking point, it must be cut into sections to reduce the stretch.

BOX AND TAPERED TAPS

The box tap (Figure 21-4) catches the outside diameter (OD) of a fish, and the tapered tap (Figure 21-5) catches the inside diameter (ID). Taps are self-threading using hardened threads and typically have vertical grooves for removing cuttings that are produced while cutting the threads. Taps have a wider, more variable catch range than an overshot or spear. However, taps are not releasable.

A special-purpose tap called a *pin tap* (Figure 21-6) is designed to catch a specific size connection, such as 3½ or 4½ internal flush (IF). Pin taps are most often used to screw into drill collars and drill pipe that have split boxes or damaged threads.

Taps should always be run with a bumper jar and safety joint because they are considered unreleasable. A drilling safety joint (Figure 21-7), the most common type used with taps, consists of a top half, friction ring, and bottom half. Spring-loaded pins keep the friction ring attached to the top half. The friction ring permits transmission of full torque in the makeup direction, but considerably less torque in the release direction.

BULLDOG OVERSHOT

The bulldog overshot (Figure 21-8) is designed for straight pickup of a fish with an OD that is too large to be caught by any other method. The overshot is a simple design incorporating a "C" grapple that fits into a tapered bowl. This design has the smallest ratio of fish OD to overshot OD of any apparatus. Like taps, the bulldog overshot cannot be released. The fish has to be free, or weaker than the work-string, so that it can be pulled in two.

FIGURE 21-4

Box tap.

(courtesy of Baker Oil Tools)

FIGURE 21-5

Taper tap.

(courtesy of Baker Oil Tools)

SPLINE JOINT

The spline joint is used in horizontal wellbore, where the need to rotate the string after the fish has been caught. As mentioned previously, it is never recommended to rotate the fishing string while tripping out of the hole for the fear of accidentally releasing from the fish. However, the spline joint is a special-purpose tool designed to provide the operator a means of rotating the drill string while engaged to the fish. Its primary use is in highly deviated or horizontal wellbores. The tool will give you the ability to transmit full torque while in the closed position but will freely rotate in the open position.

During the run-in position, the spline joint is locked closed with shear pins, and once the fish is latched, an overpull will be done to overcome the shear pin value and the tool will open to a neutral

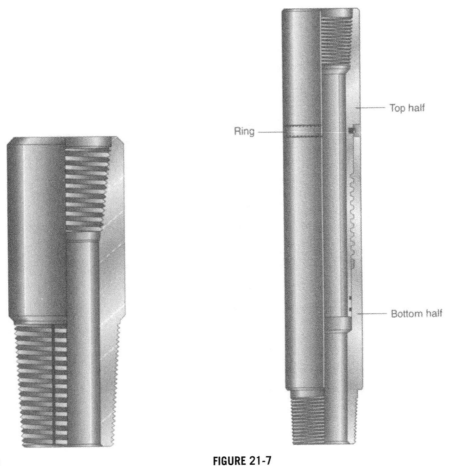

FIGURE 21-6

Pin tap.

(courtesy of Baker Oil Tools)

FIGURE 21-7

Drilling safety joint.

(courtesy of Baker Oil Tools)

position; and then the fish can either be pulled from the wellbore or jarring can start. Should there be a need to release it from the fish, the tool is lowered to close it, and engagement of the splines will occur to give the operator the ability to rotate off the fish.

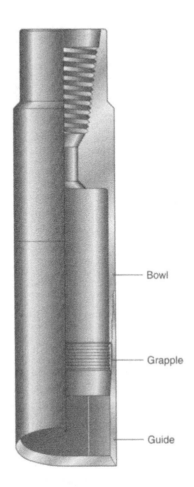

FIGURE 21-8

Bulldog overshot.

(courtesy of Baker Oil Tools)

THRU-TUBING STANDARD TOOLS

22

When running any bottom-hole assembly (BHA) on coiled tubing, a standard string of basic tools is required. In almost all cases, this consists of a coiled tubing connector, a dual-back pressure valve, a hydraulic disconnect, and (especially in mud-motor operations) a dual circulating valve. These tools are screwed together using a common set of small-diameter threaded connections (Figure 22-1).

COILED TUBING CONNECTOR

The coiled tubing connector is used to attach the BHA to the coiled tubing work-string. Available designs include a "roll-on" design, in which a rolling tool is used to deform the wall of the coiled tubing onto grooves on the connector outside diameter (OD); a set-screw design, in which screws secure the connector using preformed indentations on the coiled tubing OD; and a dimple design, in which a hydraulic tool deforms the wall of the tubing into dimples located in the connector OD. The most widely used connector is the slip-type design (Figure 22-2), which has proven superior for fishing applications, as well as for workover motor operations because of its excellent tensile and torsion properties.

The slip-type connector uses an internal gripping slip to engage the coiled tubing when making up the connection. The slip is energized using an acme thread, which provides the force required for the slip teeth to set on the outer wall of the coiled tubing. The slip-and-bowl concept used on this design allows the slip-setting force to be increased as tensile forces are applied to the joint. As more tension is applied to the coiled tubing, the joint between the coiled tubing and the connector tightens. In most cases, the connection between the coiled tubing and connector is stronger than the ultimate tensile strength of the coiled tubing itself.

The connector is rotationally locked to the coiled tubing by side-mounted set screws in the connector body. These are tightened against the coiled tubing after the slip has been set.

A pressure seal must be formed between the coiled tubing and the connector. Some tools run in the BHA are actuated by applying coiled tubing pressure after a ball has landed. These tools will be ineffective if there are any leaks in the BHA above them. The slip-type connector uses two O-ring pressure seals for redundancy and provides a positive seal with the end of the coiled tubing after the seal surface has been dressed.

Small-Diameter Tool Joint Dimension and Strength Data ▲

Joint	OD (In.)	ID (In.)	Strength (Lbs.)	Strength (Ft-Lbs.)	Makeup Torque (Ft-Lbs.
1 AMMT	1 ⁹⁄₁₆	¾	68,100	765	500
1 ¼ AM MT	1 ¾	⅞	93,984	1,265	650
1 ¼ FJ (Reg)	2 ³⁄₁₆	⅝	155,300	2,150	1,075
1 ½ AM MT	2	1	127,100	1,770	950
2 ⅜ AM PAC DSI■	2 ⅞	1 ½	201,900	6,490	3,250
2 ⅜ API Reg	3 ⅛	1	375,500	7,500	3,700
2 ⅞ AM PAC DSI■	3 ⅛	1 ½	269,470	7,290	3,650

▲ Strengths based on a material yield of 120,000 psi. Check the applicable pipe for comparative values in tension and torsion.

■ Double shoulder internal · special tri-state modification.

FIGURE 22-1

DUAL-BACK PRESSURE VALVE

The dual-back pressure valve (DBPV; shown in Figure 22-3) is basically a double-barrier isolation valve. Its purpose is primarily to prevent flow from the wellbore from coming up the coiled tubing string if the tubing ruptures at the surface. Coiled tubing failure at the surface most often happens because of bending stresses induced when running the tubing over the gooseneck and through the injector head.

The DBPV is available in at least two types. The most popular and effective type uses a double-flapper closure mechanism. The flapper design has a torsion spring to keep the valve closed if there is no flow through the coiled tubing from the surface. A bonded-rubber seat is used for pressure sealing when the flapper is closed.

The DBPV is usually run immediately below the coiled tubing connector, so the closure mechanism must allow easy passage of the largest ball used to operate the tools run below it. The largest ball used in coiled tubing operations is typically ⅞-in. in diameter. Larger balls must travel through the valve at fairly low pump rates, so spikes in coiled tubing pressure may be seen at the surface. This indicates proper tool actuation and is more difficult to recognize when using high pump rates that result in higher pressures due to fluid friction. This problem is magnified in highly deviated and horizontal wellbores, where the beneficial effects of gravity in getting the ball through are minimal.

The DBPV is also used in pressure-deployment operations such as running a wireline lubricator in a long, coiled tubing-conveyed tool string when the coiled tubing lubricator system cannot handle the length. The tools are run in the well from the wireline lubricator until a sub in the tool string, which is the same OD as the coiled tubing, is adjacent to the coiled tubing blowout preventers (BOPs) flanged to the wellhead. The coiled tubing BOPs are then closed on this sub and the wireline lubricator is bled off and rigged down. During this operation, the DBPV in the tool string prevents wellhead pressure from coming up through the coiled tubing tool string. In situations of high wellhead pressure or sour gas, the DBPV performs a critical well-control function in pressure-deployment applications.

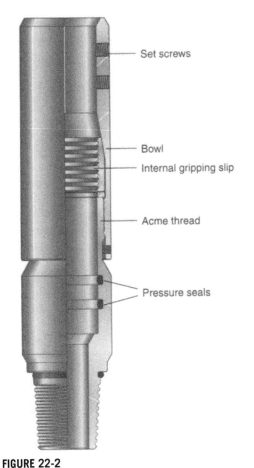

FIGURE 22-2

Slip-type coiled tubing connector.

(courtesy of Baker Oil Tools)

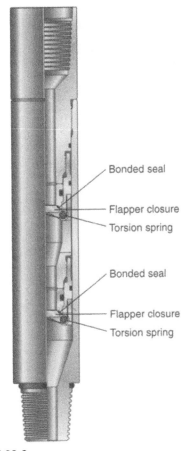

FIGURE 22-3

Dual-back pressure valve.

(courtesy of Baker Oil Tools)

HYDRAULIC DISCONNECT

The hydraulic disconnect (Figure 22-4) is used in many different coiled tubing operations including fishing and workovers. It is used to release the coiled tubing string from the BHA. This is required if the assembly becomes stuck and cannot be freed with coiled tubing tension. The hydraulic disconnect is activated by circulating a ball through the coiled tubing string and landing it in a seat within the disconnect. Hydraulic pressure will then shift a piston within the tool that disengages a locking mechanism. A slight overpull on the disconnect will allow the tool to release from the tool string below. Once the hydraulic disconnect has been activated, an internal GS-style fishneck (see Chapter 23) is exposed on the section left in the well to provide latching up during subsequent fishing operations.

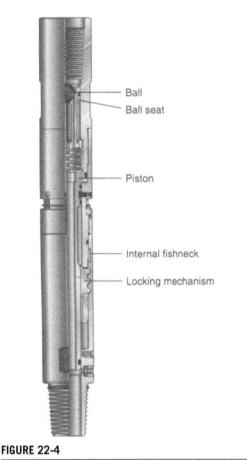

Ball

Ball seat

Piston

Internal fishneck

Locking mechanism

FIGURE 22-4

Hydraulic disconnect.

(courtesy of Baker Oil Tools)

The hydraulic disconnect is rotationally locked, so it can be used in milling, underreaming, and cutting operations on coiled tubing using a workover motor. The hydraulic disconnect cannot be run below a workover motor because the rotor-stator design of the motor will not allow passage of a drop ball to actuate the tool.

A hydraulic disconnect should be run as low in the tool string as possible—beneath jars, for example. This allows the maximum number of tools to be recovered should the assembly become stuck and the disconnect activated. If the disconnect is located above the jarring assembly, subsequent fishing operations must jar through the jarring assembly left in the hole. This would result in greatly reduced efficiency, as the lower jarring assembly would dampen the jarring load. It is possible to run one hydraulic disconnect directly below the DBPVs and another, with a smaller ball seat, below the jars.

DUAL-ACTUATED CIRCULATING VALVE

A dual-actuated circulating valve (Figure 22-5), developed mainly for coiled tubing workover-motor applications, has a traditional sleeve valve-opening mechanism with a rupture-disk port below the sleeve valve. The rupture-disk port allows circulation through the coiled tubing if the workover motor gets plugged with debris. If this occurs, pressure applied to the coiled tubing will blow out the side-mounted disk when rupture pressure has been reached.

Rupture disks are available in many different pressure ratings, and one should be selected with rupture pressure just below the safe surface limit for coiled tubing internal pressure set by the operator. Once the rupture disk is blown out, a ball can be circulated through the coiled tubing string to the sleeve valve above the rupture disk. Applying coiled tubing pressure again will open this valve and large-area flow ports.

The dual-actuated circulating valve serves three main purposes in workover-motor applications. They are as follows:

- It provides a flow path through the coiled tubing if the motor becomes plugged during operation.

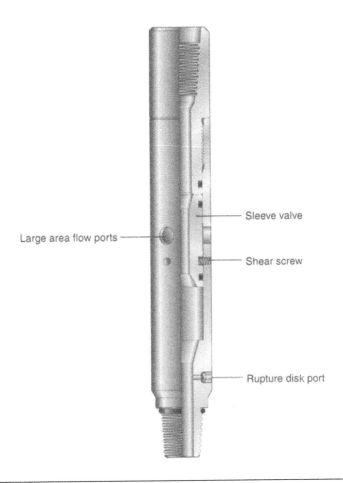

Large area flow ports —

Sleeve valve

Shear screw

Rupture disk port

FIGURE 22-5

Dual-actuated circulating valve.

(courtesy of Baker Oil Tools)

- It provides a large-area flow path out of the coiled tubing during hole cleanup after milling or underreaming. This prevents flowing through the workover motor, which creates additional back pressure and increases wear on the motor.
- It diverts fluid flow through the coiled tubing to the annulus above the workover motor when pulling out of the hole. This prevents the motor from spinning when it comes out of the hole.

MOTORHEAD ASSEMBLY

The motorhead assembly combines the main tool components described previously in this chapter, in modular form. The combined motorhead assembly is much shorter than standard BHA components because the threaded connections of the individual components are eliminated. Motorhead assemblies offer easier handling, take up less lubricator height, and reduce bending stresses in the BHA.

THRU-TUBING FISHING

23

PREPLANNING

Fishing with coiled tubing requires extensive preplanning that involves all participating personnel, including the operator, representatives from the service companies, and representatives from the fishing-tool company. It is advisable to review service reports and consult with personnel who have been involved in previous fishing operations. A thorough analysis of the well's history and events leading to the need for a fishing operation will save time on determining the location and equipment requirements. Detailed information should include (but are not limited to) the following:

- The most recent well schematic, including minimum restrictions
- The available riser height (to determine if there is enough space to recover the fishing assembly and fish)
- Data about coiled tubing forces, particularly overpull, set-down weight, and the fatigue-cycle life of the reel
- The tensile strength of the tools, particularly those run below jars

Take the time needed to develop the best strategy and contingency plans. Considerations such as equipment availability, logistics, methodology, operator constraints (especially the cost and the time allocated), and engineering and manufacturing support all should be addressed during preplanning. This stage is critical to development of the procedure that offers the maximum chance of success.

RULES OF THUMB

In order to minimize risk when fishing with coiled tubing, the following rules of thumb should be considered:

- Avoid running tools that cannot be externally fished in the hole
- Review the outside diameter (OD) of the overshot to ensure that it will pass through the minimum restriction in the well; clearance through minimum restriction should be at least $\frac{1}{16}$ in.
- Centralize the bottom-hole assembly (BHA) to reduce drag (assuming that the fish is centralized). This will also leave a centralized fishneck should the BHA have to be disconnected.
- Function-test hydraulic-actuated tools at the surface to determine the flow rates at which they operate. This is especially important when using flow-release overshots and spears.

159

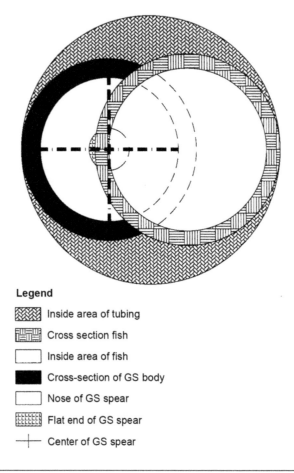

Legend

▨ Inside area of tubing

▨ Cross section fish

☐ Inside area of fish

■ Cross-section of GS body

☐ Nose of GS spear

▤ Flat end of GS spear

—┼— Center of GS spear

FIGURE 23-1

Layout sketch of fishing geometry.

- Create a layout sketch of fish and fishing tool geometry in worst-case positions in the wellbore (Figure 23-1). Actual-size sketches are most effective in helping to visualize potential problems during latchup.

ADDITIONAL EQUIPMENT

In addition to the coiled tubing unit and crew, equipment involved in a fishing operation may include a wireline unit, fluid mixing and pumping equipment, and nitrogen equipment.

Whenever possible, take advantage of wireline trip speed, wireline sensitivity, and crew experience. A wireline unit and crew are important parts of a fishing operation. They can perform many jobs, such as impression-block runs, retrieving light obstructions, and "baiting" the fish to facilitate coiled-tubing tool engagement, and using wireline can provide deployment and retrieval assistance more efficiently than using coiled tubing. When down-hole conditions are uncertain, wireline can be used to examine alterations in downhole conditions or fish configuration. Investigative trips with coiled tubing are time-consuming and often unproductive compared to using wireline. The equipment on location should be placed to facilitate quick rig-up/rig-down of the coiled tubing and wireline equipment.

Pumping equipment is essential. Fluid circulation can be used to wash debris from the fish, activate fishing tools, and displace balls to activate various tool-string components. Mixing equipment may be required for the preparation of friction reducers or weight additives. Friction reducers introduced into the wellbore can increase impact force and facilitate retrieval. Kill-weight fluids may be required for controlling the well, altering hydrostatic pressure, or increasing the buoyancy of the coiled tubing.

Nitrogen equipment on location can also be useful during the fishing operation. Displacing the coiled tubing with nitrogen will increase buoyancy and residual overpull at depth. Nitrogen can also be introduced into the wellbore to lighten hydrostatic pressure exerted against the fish.

SPEARS AND OVERSHOTS

Most types of conventional spears and overshots can be conveyed on coiled tubing. These include standard basket- and spiral-type overshots, Kelo sockets, mouse-trap overshots, and releasing spears. However, these types of tools cannot be released conventionally because coiled tubing cannot be rotated. If these tools are run on coiled tubing and the fish cannot be retrieved after latching the overshot, a hydraulic disconnect farther up the tool string must be activated, which leaves additional tools in the hole. To avoid this situation, run hydraulic-release spears and overshots before running conventional spears and overshots with coiled tubing.

HYDRAULIC-RELEASE SPEARS AND OVERSHOTS

Hydraulic-release spears (Figure 23-2) and overshots (Figure 23-3) are designed specifically for coiled tubing fishing operations. The overshot is used to catch either external fishnecks or slick ODs, and the spear is used in internal fishnecks or slick inside diameters (IDs). Hydraulic-release overshots and spears for fishneck profiles are dressed with a collet designed to fit a specific type of fishneck (as shown and specified in Figures 23-4 and 23-5). Overshots and spears designed to catch slick fishnecks are dressed with a grapple-type collet in the specific catch size range required.

These tools engage the fish with only a little coiled tubing slackoff weight, making them useful in highly deviated or horizontal applications, in which set-down weight is limited. The spear or overshot is disconnected from the fish by circulating through the coiled tubing. No rotational movement is required. Circulation passes through a removable, adjustable orifice in the bottom of the tool. The orifice creates back pressure in the tool, which actuates a piston that releases the catching mechanism. The spear or overshot can then be pulled off the fish. The releasing mechanism does not shear screws or any other positioning devices, so the tool may be relatched to the fish as required without a trip out of the hole. The hydraulic release allows the entire fishing string to be pulled from the well if the fish cannot be pulled.

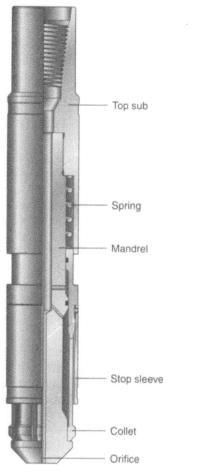

— Top sub

— Spring

— Mandrel

— Stop sleeve

— Collet

— Orifice

FIGURE 23-2

Hydraulic-release spear.

(courtesy of Baker Oil Tools)

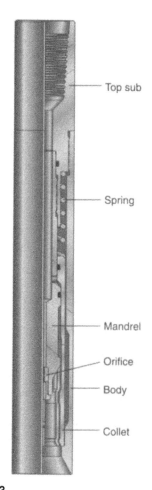

— Top sub

— Spring

— Mandrel

— Orifice

— Body

— Collet

FIGURE 23-3

Hydraulic-release overshot.

(courtesy of Baker Oil Tools)

Another advantage of hydraulic-release spears and overshots is their ability to circulate fluid out of the end of the tool. This provides a means of washing debris or filling out the fishneck to allow proper latching of the spear or overshot.

Both tool designs use a collet-style grapple to latch the fishneck. The collet itself does not get loaded during the jarring operation. The grapple section that catches the fish is in compression when pulling or jarring on the fish.

FISHING IN LARGER DIAMETER BORES

Attempting to locate and latch small fishnecks in larger diameter bores using coiled tubing can often prove challenging. Tools such as hydraulic centralizers and knuckle joints can be used to help latch the fish with the spear or overshot.

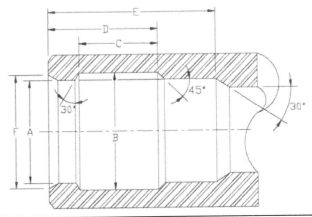

FIGURE 23-4

Internal fishing necks.

(courtesy of Baker Oil Tools)

Internal Fishing Necks									
Size	A (In.)		B (In.)		C (In.)	D (In.)	E (In.)	F (In.)	
	Max	Min	Max	Min				Max	Min
1¼	.90	.88	1.05	1.03	1	1.38	2	1.03	1.00
1½	1.08	1.06	1.24	1.22				1.19	1.16
2	1.40	1.38	1.58	1.56				1.62	1.59
2½	1.83	1.81	1.99	1.97				1.94	1.91
3	2.33	2.31	2.52	2.50	1½	2	3	2.50	2.47
3½	2.64	2.62	2.83	2.81				2.81	2.78
4	3.14	3.12	3.33	3.31				3.38	3.35
5	4.02	4.00	4.21	4.19				4.19	4.16
7	5.38	5.40	5.62	5.64				5.60	5.64

In cases where the fishneck is away from the center of the wellbore, an indexing tool can be run in combination with a hydraulic bent sub (Figure 23-6) to systematically search the larger bore and engage the fish. Flow through the indexing tool rotates the tools run below it in 30° increments. The indexing tool is actuated using back pressure created from the orifice located in hydraulic retrieving tools run below it. The hydraulic bent sub provides the means of deflecting the fishing tool at an angle of between 2° and 10°.

FISHING PARTED COILED TUBING

When coiled tubing parts under tension, it necks down over several inches at the point where it parted. It is possible for overshots to engage the top of the parted section, but engaging parted coiled tubing can be difficult because of residual helix and spring in the coil itself.

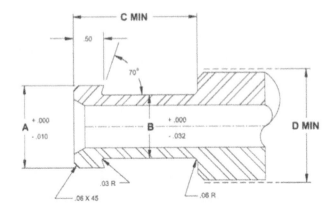

External Fishing Necks						
Min Tubing Size in Which Neck May Be Run (In.)	**A▲ (In.)**	**B (In.)**	**C▪ (In.)**	**D● (In.)**	**Pulling Tool**	
					Otis	**Camco**
1.660	0.875	0.688	2¾	1 5/16	1 3/16 RB	1-1/4 JDC*
					1 5/16 SM*	1-5/16 JDC
1.660	1.000	.813	2¾	1 7/16	1¼ RB	1-3/8 JDC*
					1½ RB*	
1.900	1.188	1.000	2¼	1½	1½ RB*	1-1/2 JDC*
					1½ SB	1-1/2 JUC
2 3/8	1.375	1.188	2 3/8	1 15/16	2 RB*	2 JDC*
					2 SB	2 JUC
2 7/8	1.750	1.500	2¼	2 3/8	2½ RB*	2-1/2 JDC*
					2½ SB	2-1/2 JUC
3½	2.313	2.063	2¼	2 7/8	3 RB*	3 JDC*
					3 SB	3 JUC

* These dimensions are based on using the pulling tools (overshot most commonly found on wire line service trucks They are the Otis RB, RS, SB, AND SS, and the Camco JDC, JUC, JDS, and JUS. The RB, RS. JUC, and JUS are all jar-up release tools, and the others are jar-down release.

▪ The minimum recommended distance to allow an Otis SB or a Camco JDC pulling tool to release.

▲ Fishing Neck Type Size sometimes referred to by "A" dimension.

● The least recommended diameter immediately surrounding the fishing neck to allow the overshot dogs room to latch and unlatch.

FIGURE 23-5

External fishing necks.

(courtesy of Baker Oil Tools)

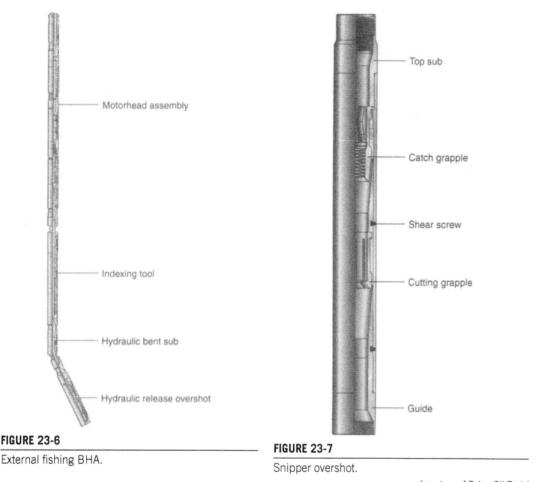

FIGURE 23-6

External fishing BHA.

FIGURE 23-7

Snipper overshot.

(courtesy of Baker Oil Tools)

When coiled tubing parts under compression, it typically buckles and bends over to form the shape of a shepherd's hook. A compression failure normally requires dressing the coiled tubing using a wash-over shoe before running tools such as snippers or continuous overshots.

SNIPPER OVERSHOT

A snipper overshot (Figure 23-7) is used to remove the damaged top section of parted coiled tubing remaining in the hole. The snipper overshot contains a catch-and-cutting mechanism that washes over the top of the coiled tubing. Once the required amount of coiled tubing (typically 5–10 ft.) is swallowed, overpull is applied, which forces the catch grapple to bite into the coiled tubing. This overpull may be enough to free the fish. If not, it is increased to activate the lower cutting grapple, which cuts the coiled tubing. It is advisable to bench-test the snipper overshot on a sample section of the coiled tubing to be fished. This indicates the overpull that will be required during the actual fishing operation. Once the snipper has been run, a round fishneck with an unrestricted ID remains, which allows passage of a chemical cutter should it be required later in the fishing operation.

CONTINUOUS-TUBING OVERSHOT AND HIGH-PRESSURE PACKOFF

The continuous tubing overshot (CTO) (Figure 23-8) and high-pressure packoff (HPP) (Figure 23-9) are used to catch coiled tubing that has parted in the hole. The CTO uses a grapple that is sized to the coiled tubing being fished. The grapple segments are arranged in a circle to provide a nearly complete slip bite. They are spring-loaded against the coiled tubing, so long sections of tubing can be washed over without dragging the grapple teeth against the tubing, which causes tooth damage. The CTO moves freely downward over the coiled tubing. The grapple segments engage with any upward

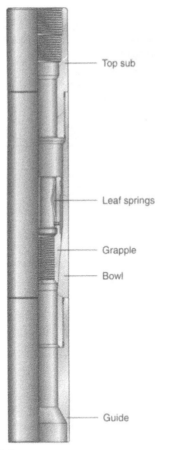

Top sub

Leaf springs

Grapple

Bowl

Guide

FIGURE 23-8

Continuous-tubing overshot.

(courtesy of Baker Oil Tools)

FIGURE 23-9

High-pressure pack-off.

(courtesy of Baker Oil Tools)

movement of the tool. Unlike the hydraulic-release overshot discussed previously, the CTO cannot be released once engaged. However, it can be run with either threaded-tubing or coiled tubing work-strings when fishing coiled tubing.

The HPP is run above the CTO and provides a high-pressure seal between the coiled tubing being fished and the work-string. This is done to circulate out around the coiled tubing being fished, which can help to free it. The HPP can also allow drop balls to be circulated to seat in circulation subs and hydraulic disconnect tools in the original BHA. Actuation of these tools can also help to free the stuck tubing.

THRU-TUBING FISHING JARS AND ACCELERATORS

24

JARS

In fishing operations, hydraulic jars are used to deliver an impact force to free the fish. Jars designed specifically for coiled tubing are generally flow-through and torque-transmitting. They are available in single-direction and bidirectional versions. The most commonly used jars are the hydraulic, time-delayed type (Figure 24-1). This jar does not rely on stored energy within the tool itself to regulate the force of impact. Instead, the hydraulic jar uses a fluid-metering system to regulate fluid transfer between the cavities of the tool when tension is applied by the coiled tubing.

The amount of time delay for jar release varies with the amount of overpull applied. Once a sufficient amount of fluid is metered through the jar, the hammer is free to move rapidly upward to strike the anvil of the tool. This upward impact is transmitted through the jar to the fish below it.

The impact force can be varied during a single run. Jarring can start with lower-impact loads that can be increased gradually if the fish does not release. This is good practice because it may free the fish with lower-impact forces, reducing the chance of damaging it. The jar is recocked by applying set-down weight with the coiled tubing.

Dual-acting jars provide the flexibility to impart downward impact as well during fishing operations. However, downward jarring on coiled tubing is less effective than upward jarring because of the lower available set-down weights and higher frictional forces created by the helical motion of the coiled tubing as it is run into the hole.

Flow-through weight bars are used directly above the jar. Weight bars convert the impact force generated by the jar alone into an impulse force and increase the magnitude of the force delivered. In thru-tubing fishing operations, the amount of weight bar run above the jar is dictated by the amount of lubricator available for the specific application.

ACCELERATORS

Accelerators are run directly above the weight bars. An accelerator (Figure 24-2) conserves the maximum amount of momentum created in the jarring assembly at jar release. It has a stroking assembly similar to that of the jar. As tension is applied to the coiled tubing to release the jar, the accelerator is also stroked. When the jar releases, the stroke in the accelerator allows the upper section of the jar, the weight bars, and the bottom section of the accelerator to move independent of the coiled-tubing workstring. Momentum-reducing drag and dampening effects of the coiled tubing are removed, allowing the jarring assembly to work at maximum efficiency.

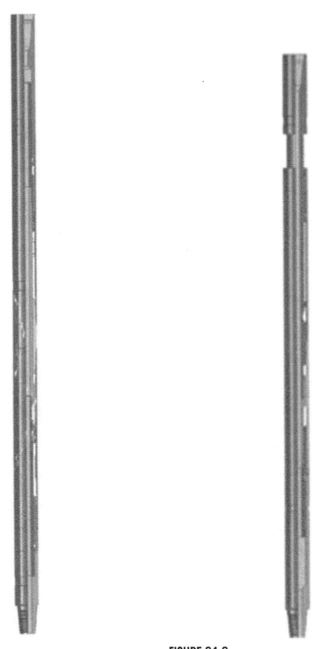

FIGURE 24-1

Hydraulic jar.

(courtesy of National Oilwell)

FIGURE 24-2

Accelerator jar.

(courtesy of National Oilwell)

When setting up a jarring assembly, the stroke created by the overpull within the accelerator to release the jars must be greater than the stroke of the jar during release. If the accelerator stroke is less than the jar stroke, the impact force will be directed upward instead of down to the fish below it. Like jars, accelerators are recocked with set-down weight applied with coiled tubing.

Hydraulic and gas-charged accelerators are available, and both types have operational considerations. Hydraulic accelerators provide considerably less impact than gas-charged accelerators. Gas-charged accelerators have to be set up for a predetermined overpull and can be less reliable than hydraulic accelerators over longer periods of use.

BIDIRECTIONAL VIBRATORY JARS

When running jars on coiled tubing, the pipe has to be cycled over the gooseneck many times to fire and recock the jar, whether in the up or down mode. Repeatedly cycling the coiled tubing can result in fatigue of the pipe section that is traveling up and down over the gooseneck. This can often be a limiting factor in the fishing job.

In some cases, the low-frequency, high-impact blows of a jarring tool can also be a limitation. A sand-stuck fish is often wedged tighter after the combination of hard jarring with small particles. If the sand particles are temporarily suspended or liquified by lighter, high-frequency blows while overpull is applied, there is a much greater chance of freeing the fish.

A bidirectional vibratory jar (Figure 24-3) can deliver both upward and downward impacts without cycling the coiled tubing. The tool is powered by fluid pumped through the coiled tubing workstring and is designed to deliver downward blows in compression and upward blows in tension.

FIGURE 24-3

Bidirectional vibratory jar.

(courtesy of Baker Oil Tools)

Downward blows are amplified and accelerated by stored internal energy. A small amount of tension is enough to cock the tool—usually about 1,500–2,000 lbs. above the pipe weight in the up mode. The tool does not contain any mechanical latches and has only three moving parts. Blow counts vary in either configuration depending on fluid volume. At high volume in the up mode, the tool acts as a vibratory extractor and has proven highly effective at extracting sand-compacted fish.

Bidirectional vibratory jars are typically used for shifting sliding sleeves, fishing, swaging collapsed tubing, breaking knockout isolation valve (KOIV) disks, and setting and retrieving wireline tools in deviated wells. The tool may be run with or without an intensifier, but it is always advisable to run an intensifier when surface testing is required, when working depths are less than 600 ft., and when tubing with a 1¾-in. or larger OD is being used.

THRU-TUBING DEBRIS CATCHERS

25

VENTURI JET JUNK BASKET

The venturi jet junk basket is used to remove various types of high-density debris and formation particles that are too heavy to be circulated to the surface. The flow in the top of the tool passes to the outside via adjustable jets. The jets cause a pressure drop inside the venturi basket, which acts as a suction at the bottom inlet of the tool. The fill is stirred up by the flow down the outside of the tool. At the suction inlet on the bottom of the tool, the fill is carried through finger cages into the tool's internal filter screen (Figure 25-1). The strained fluid then passes through the top of the tool and mixes with the pumped fluid. Fill is trapped between the filter and the finger cages. Extension cylinders can be added between the filter and finger cages to increase the amount of fill that the tool can carry out of the hole.

Venturi jet junk baskets can be fitted with a dressed rotary shoe and run below a mud motor to break up and retrieve debris compacted in the wellbore or around a fishneck (Figure 25-2).

MAGNETIC CHIP CATCHER

On a milling job, a magnetic chip catcher (Figure 25-3) can be run above the motor to help remove cuttings from the well. The tool contains magnets that attract metal chips from the wellbore fluid. This tool is ideal for use on coiled tubing, where annular fluid velocities are too low to carry cuttings out of the well. The catcher is usually run directly above the motor. A centralizer is incorporated into the tool to stand it off the casing so cuttings will not be scraped off the tool while it is pulled out of the hole.

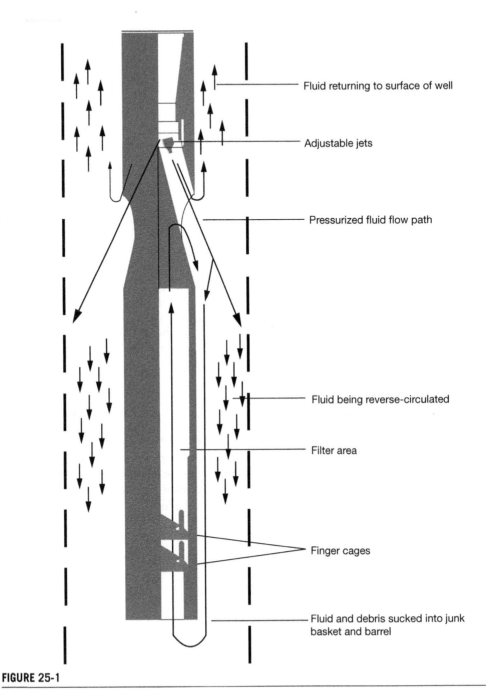

Fluid returning to surface of well

Adjustable jets

Pressurized fluid flow path

Fluid being reverse-circulated

Filter area

Finger cages

Fluid and debris sucked into junk basket and barrel

FIGURE 25-1

Venturi jet junk basket showing a flow-path direction.

(courtesy of Baker Oil Tools)

Motorhead assembly

Workover motor

Venturi jet basket

FIGURE 25-2

Rotary venturi cleanout bottomhole assembly.

(courtesy of Baker Oil Tools)

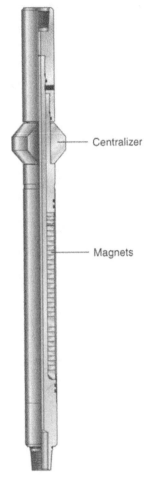

Centralizer

Magnets

FIGURE 25-3

Magnetic chip catcher.

(courtesy of Baker Oil Tools)

THRU-TUBING WORKOVER MOTORS

Thru-tubing workover motors are run through production tubing and provide rotation and torque for coiled tubing fishing, underreaming, milling, and cutting.

WORKOVER MOTOR COMPONENTS

Positive displacement motors (PDMs) have three major elements (Figure 26-1). From top to bottom, they are the power section, transmission, and output shaft and bearing assembly.

POWER SECTION

The power section of the motor converts the hydraulic energy of the drilling fluid into mechanical horsepower to drive the bit. The mechanical horsepower of a multilobe, rotor-stator power section is a product of high-output torque and slow rotational speed. Typically, a power section consists of a steel rotor and an elastomeric stator.

The elastomeric stator is bonded to a tubular-steel housing (Figure 26-2). The multilobe form is helically screwed along the length of the stator. The steel rotor is produced with matching lobe profiles and a helical pitch similar to the stator, but with one less lobe. This allows the rotor to be matched and inserted into the stator. Drilling-fluid flow creates hydraulic pressure that causes the rotor to precess, as well as rotate, within the stator. This type of movement is called nutation.

New power-section designs use a stator with an elastomer layer of uniform thickness instead of the conventional, variable-thickness stator. This type of stator, known as an *equidistant* stator (Figure 26-3), has a machined-steel internal profile that requires much less rubber compound. Because its lobes can withstand greater pressure, the equidistant power section can generate a higher operating torque. Having less rubber allows the tool to operate in higher temperatures than with the conventional stator.

TRANSMISSION

The transmission assembly transmits the rotational speed and torque produced by the power section to the output shaft. It must be capable of absorbing the downward thrust generated by the power section so that the rotor stays in correct axial alignment with the stator. The transmission must also eliminate processional motion of the rotor and deliver only concentric rotational speed to the output shaft.

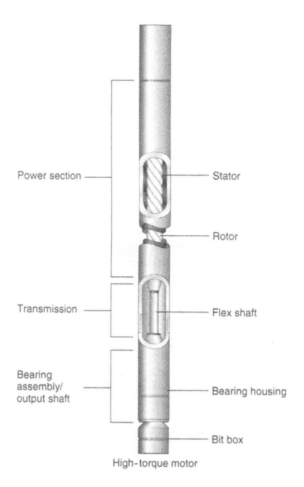

Power section —

Stator

Rotor

Transmission —

Flex shaft

Bearing
assembly/
output shaft

Bearing housing

Bit box

High-torque motor

FIGURE 26-1

Positive-displacement motor.

(courtesy of Baker Hughes).

FIGURE 26-2

Elastomeric stator.

(courtesy of Baker Hughes)

FIGURE 26-3

Equidistant stator.

(courtesy of Baker Hughes)

OUTPUT SHAFT AND BEARING ASSEMBLY

Most of the workover motor's drilling fluid passes through the center of the output shaft to the drill bit. This shaft is constructed of rigid, hollow steel and is supported within the bearing casing by radial and thrust bearings.

Thrust bearings accept downward thrust from the rotor and reactive force from weight on the bit, which thrusts upward. Thrust bearings are multistack, ball-track types. Radial bearings are hardened, lined sleeves running on a hardened surface.

A sleeve flow restrictor is mounted below the upper radial bearing to reduce the amount of drilling fluid passing through the bearing assembly. The small amount of fluid passing through the flow restrictor lubricates and cools the bearing assembly. A bit-box, threaded connection couples directly to the output shaft and screws onto the bit.

WORKOVER MOTOR PERFORMANCE

The PDM workover motor produces torque as fluid is pumped through its power section. The rotor-stator configuration determines the motor's flow rate, speed, differential pressure, and torque characteristics.

Fluid in the motor is channeled through the power section, following the cavity created by the rotor, which has one less lobe than the stator. The lobe configuration, patterned in a helix, provides a sealed chamber for the fluid. This chamber is created because the helical pattern of the stator is longer than the rotor. One full revolution of the stator lobe helix constitutes a stage, and each stage within a power section provides additional torque. Correspondingly, the pressure differential increases with the number of stages.

Multilobe, rotor-stator configurations are designed to provide a working range of torque, length, and speed to suit coiled tubing operations. The multilobe configuration acts as a gear reducer to provide higher torque at reduced speed, much like a planetary and sun gear arrangement.

Based on the rotor-stator configuration, the motor's speed will correspond to the flow rate until it is restricted, as with weight on the bit. As weight is applied to the bit, this torque demand will be seen as differential pressure (Figure 26-4).

If the weight on the bit continues to increase, the motor's speed will decrease to the point of stalling, which occurs when the resistance to rotation produced at the bit overcomes the sealing capacity between the rotor and stator. At this point, the flow bypasses the normal path through the power section.

OPERATIONAL CONSIDERATIONS

Before using a workover motor on coiled tubing applications, calculations must be made to determine if enough hydraulic power is available down-hole. This is especially important in deep wells in which a long coiled tubing string of a shorter length is used because fluid friction becomes a larger factor. Three components to be considered in calculating thru-tubing workover-motor hydraulics are pressure loss in the coiled tubing, pressure loss across the bottom-hole assembly, and pressure loss in the annulus.

The chemical composition of both fluids pumped from the surface and also aggressive wellbore fluids should be considered. Many fluid types can seriously damage the stator elastomer. A fluid/elas-

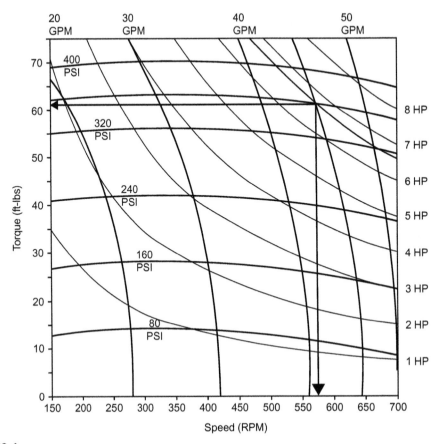

FIGURE 26-4

Graph showing the performance of a 1¹¹⁄₁₆-in. motor at 45 gpm and 360 psi differential pressure (on/off bottom). Plot constant-flow line for 45 gpm and constant-pressure line for 360 psi. From the point at which these two lines meet, a straight line to the left margin shows a torque output of 61 lb-ft., and a straight line down shows a bit speed of 570 rpm. Following the curve of constant power to the right gives 6.7 hp to the engine.

(courtesy of Baker Hughes)

tomer reactivity database, obtained from the workover motor provider, should be consulted before the job. This information contains elastomer reactivity information for oilfield chemicals, which can help prevent job failure because of elastomer damage. If wellbore fluids are known to be aggressive, lab testing for elastomer reactivity should also be performed in advance of the job.

Well temperature should be considered and checked with the temperature rating of the workover motor. The effect of high temperatures is to swell the stator elastomer, which in effect makes the inside diameter of the stator smaller and increases interference between the rotor and the stator. This could have an adverse effect on the performance of the motor. Conventional power sections are rated to perform well at temperatures up to 320 °F, with equidistant stators rated as high as 400 °F.

Torque is stored in the coiled tubing when the workover motor stalls. If the motor is lifted off bottom without releasing this torque, the whipping action of the tubing can back off a tool-string connection or cause torsional damage to the coiled tubing. Stored torque can be released safely by stopping the pumps and allowing pressure in the coiled tubing to bleed off before pulling it and the bottom-hole assembly (BHA) up-hole, away from the stall point.

Pressure-actuated tools can also be used above the mud motor. They function on back pressure created by fluids pumped through the motor and are adjustable for different actuating pressures. The amount of expected back pressure must be known for proper tool setting.

USING COMPRESSIBLE FLUIDS

When operating positive-displacement workover motors with compressible fluids, these guidelines will increase the chances of success:

- As a rule of thumb, when a compressible fluid such as nitrogen is used to drive the motor, output torque will decrease by approximately 30%. A motor should be run with enough torque-output capacity to account for the difference.
- Conversion factors should be used to correlate the correct standard cubic feet (scf) per minute flow rate of nitrogen to an equivalent flow rate in gallons per minute (gpm). These conversion factors vary depending on the job. Software modeling programs are available to assist in providing more accurate operating information in these conditions.
- When rigging up and performing the surface test, be sure to supply the correct amount of nitrogen and fluid to the motor.
- The following factors also should be considered when drilling with nitrogen-based fluids:
  - Introduction of nitrogen gas into the stator rubber may cause blistering upon return to surface (explosive decompression) and render the stator unfit for future runs.
  - The motor is more sensitive to weight on the bit with nitrogen-based fluids than with conventional drilling fluids.
  - Dry nitrogen should not be pumped.
  - The motor stalls at lower differential pressures than with conventional drilling fluids.
  - Because of the underbalanced conditions, less weight on the bit is required to drill.
  - Motor stall is difficult to see on the surface because of the compressibility of nitrogen drilling fluid.

THRU-TUBING MILLING

27

As recently as the early 1990s, coiled tubing milling was considered a high-risk operation that often left broken and stuck tools in the hole. Poor job planning, unreliable motor performance, and inexperienced personnel contributed to the failure rate, and tools in the milling bottom-hole assembly (BHA) often failed to withstand torque and vibration stresses from the motor.

Today, coiled tubing milling is considered fairly routine, with a high overall success rate. Most of the previous problems have been overcome, and the versatility and potential of coiled tubing milling operations are now being realized.

Coiled tubing is used to mill materials such as scale, metal, cement, composite, and cast-iron bridge plugs, as well as many forms of loose junk. When milling metal, computer software can be used to calculate appropriate mill-bit speeds to provide optimum milling performance (Figure 27-1).

OPERATIONAL PROCEDURES

Once a decision is made to mill and the operating conditions have been defined, procedures can be designed. The following steps are recommended in typical operations:

- Inspect all tools.
- Take measurements and drift inside diameters (IDs).
- Attach coiled tubing connector and perform pull test.
- Make up standard running tools to circulation sub and pressure test.
- Make up the workover motor and surface-test it, recording flow rates and pressures throughout the operating range.
- Make up the mill bit, pull complete BHA into lubricator, and pressure-test the lubricator.
- Run into the hole at a controlled speed, slowing through completion accessories.
- Establish the starting depth by lightly tagging the object to be milled.
- Pick up approximately 50 ft. and establish the milling flow rate and circulating pressure.
- Run into the hole until circulation pressure increases.

Note: At this point, patience is critical. Motor stalls can occur repeatedly until a milling pattern is established. The set-down weight should be applied until approximately a 300–400 psi differential above the off-bottom circulating pressure is maintained.

- Pump the gel sweeps as dictated by annular velocities to circulate out cuttings.
- Once the target depth is reached, perform verification passes across the milled area.
- Activate the dual-circulation sub and pump at the maximum rate while pulling out of the hole.

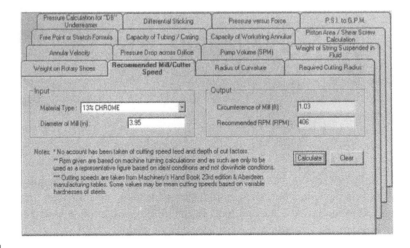

FIGURE 27-1

Milling software program.

(courtesy of Baker Oil Tools)

SCALE MILLING

Scale milling and profile enlargement are two of the most frequently performed coiled tubing milling operations. Scale buildup of solid deposits is a common occurrence in tubulars and casing. Scale deposits can restrict or totally block the flow area, resulting in a partial or total loss of production. They must be efficiently removed from the tubing and casing to return production to normal levels. These deposits typically occur where oil, reservoir, and injection water is processed. Problems usually arise with changes in pressure, temperature, and velocity. These scale deposits listed next are commonly found throughout the oilfield and range from relatively soft to extremely hard, crystallized structures:

ORGANIC

Paraffin (wax)
Asphalt

INORGANIC

Calcium carbonate
Calcium sulphate
Iron carbonate
Iron sulphide
Iron oxide
Strontium sulfate
Barium sulfate

Mechanical scale-removal methods work well, particularly in removing harder types of scale. Impact drilling can provide a viable solution, but the most frequently used scale-removal method is milling. Constant torque applied to the mill bit by more powerful and reliable workover motors is a significant reason for the effectiveness of this method, along with advances in mill-bit cutting inserts.

A typical milling assembly (Figure 27-2) includes standard running tools, a nonrotating stabilizer, a workover motor, and a mill bit. A nonrotating stabilizer is included because milling performance has proven better with the assembly stabilized above the workover motor. The stabilizer OD should be $\frac{1}{16}$ in. less than the mill-bit OD. This provides optimum stabilization of the assembly and prevents sticking problems should the mill bit become worn.

Many types of mill designs can be used to mill scale. Cutting matrixes, including diamonds, polycrystalline diamond compact (PDC), and tungsten carbide, have all been used successfully. The most widely used matrix consists of a very aggressive, tungsten carbide milling matrix formed in the shape of a dome (Figure 27-3). This milling-head design presents a small contact area to the scale, which reduces torque during milling. Although this mill has an aggressive face, it does not damage tubing walls. Specially designed stabilization pads directly behind the milling matrix on the outside diameter (OD) of the mill body provide the required protection.

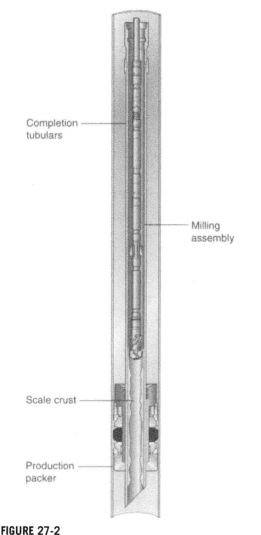

FIGURE 27-2

Typical milling assembly.

(courtesy of Baker Oil Tools)

FIGURE 27-3

Scale-removal mill.

(courtesy of Baker Oil Tools)

PROFILE ENLARGEMENT

Enlarging or removing nipple profiles is usually performed during thru-tubing operations where the ID of the nipple is less than the OD of the tools required to pass through it. In many cases, the lower-tailpipe nipple of the completion is removed so contingency fishing tools have access to the liner below it. A clearance of 1/16 in. is required between the OD of the fishing tools and the ID of the minimum restriction through which they need to pass.

A profile-enlargement milling assembly includes the same BHA configuration as the scale-milling assembly, but with a different mill-bit design (Figure 27-4). The profile-enlarging mill consists of a series of steps with a wear pad at the top of the largest OD step. Each step is dressed with tungsten carbide cutting inserts, and the length of each step is custom-built to suit the dimensions of the nipple being enlarged. The step feature is designed to keep torque consumption of the workover motor low and cutting sizes to a minimum. This mill design is based in part on lathe-cutting principles and leaves a machinelike finish on the milled surface.

FIGURE 27-4

Profile enlarging mill.

(courtesy of Baker Oil Tools)

THRU-TUBING UNDERREAMING

28

An underreamer is designed to pass through a down-hole restriction, open up below the restriction, clean the hole to full gauge, and then close up for retrieval through the restriction. The restriction is typically in the production-tubing string in the form of nipple profiles, mandrels, and other completion accessories.

The most common underreaming task is removing cement left from coiled-tubing squeeze cementing. Leftover cement is typically the result of large cement nodes forming at the squeezed perforations or cement hardening before the excess can be reversed out. These cement restrictions must be removed before re-perforating can be done. The underreamer is also used to clean out scale and hard fill that cannot be removed from liners by jet-washing tools.

A mill could be used in these conditions, but the resulting hole size would be about the same as the internal drift of the tubing, leaving a sheath on the walls of the liner. This sheath could dislodge during subsequent operations, possibly resulting in stuck tools. Perforating would be less efficient because the charges would have to expend energy penetrating the sheath before they reached the liner.

Thru-tubing underreamers are available in two-bladed or three-bladed designs. The three-bladed tool can also be used as a pipe cutter by replacing the underreaming knives with cutting blades. A detailed discussion of the three-bladed underreamer (also known as a *hydromechanical pipe cutter*) is included in Chapter 29. Two-bladed designs are more common in the smaller tools [typically with 2¼ in. outside diameter (OD) and under (Figure 28-1)]. Many underreamers have something called a *bit box*, which allows a mill to be screwed to the bottom of the tool. The added mill reduces workload on the underreamer knives and can eliminate a separate milling trip if a long, solid cement plug has to be removed.

Several tools are run in combination with the underreamer. Some tools (for example, the workover motor) are required, while others are discretionary depending on the particular application and procedure. Once underreaming is complete, it is usually desirable to increase the pump rate to clean the hole. To do this, the motor and underreamer must be bypassed because of the pressure loss across these tools. Opening the dual-circulation main ports provides the ability to do this.

OPERATIONAL CONSIDERATIONS

The-following guidelines have proven useful in most situations for preparing and executing a successful underreaming job with coiled tubing.

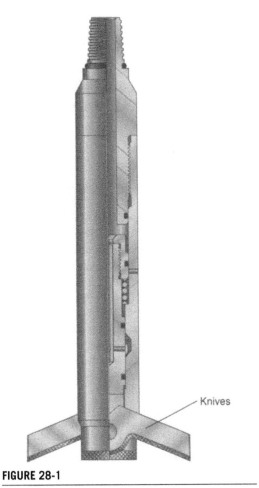

Knives

FIGURE 28-1

Two-blade underreamer.

(courtesy of Baker Oil Tools)

BEFORE RUNNING IN

- Once the coiled-tubing equipment is rigged up, the underreaming assembly is made up and typically consists of a coiled-tubing connector, dual-flapper check valves, a hydraulic disconnect, a dual-circulation sub, a workover motor, and an underreamer.
- To minimize the potential for leaks or backoffs during underreaming, three precautions should be taken:
  1. Each part of the assembly should be threadlocked and made up with the correct torque. At high rotating speeds, spinoffs can occur if connections are not made properly.
  2. Crossover subs should be kept to an absolute minimum.
  3. All tools should be pressure tested correctly.
- While picking up the bottom-hole assembly (BHA), tools should be made up down to the circulating sub and pressure-tested through the coiled tubing before picking up the motor and underreamer. This step ensures connection-pressure integrity. After installing the workover motor and underreamer, the BHA should be function-tested at the surface. The flow rate at which the underreamer opens and the motor begins to turn should be noted.
- Once the tool string is pressure- and function-tested, the lubricator should be made up and pressure-tested. It is preferable to perform the lubricator pressure test by pumping through the pump-in sub instead of through the coiled tubing. Pumping through the coiled tubing may open the underreamer's arms and cause it to hang up in the tree or other restriction in the well.

AFTER RUNNING IN

- When opening the underreamer, the arms should not be constrained. Opening the tool in a sheath may damage it or prevent it from opening. This can also result in drilling a pilot hole instead of a full-gauge hole. To avoid this, pull the underreamer to a point at which there is no restriction. Once the underreamer is in position, initiate pumping to open the tool.
- Running speed should be kept below 30 ft./min when underreaming in the liner from the starting point down to the restriction or fill. This allows the underreamer to act as a casing scraper. When

the first significant restriction is encountered, the pump pressure will increase and the weight indicator will show a loss of weight.

- When a restriction is encountered, allow the underreamer to set a pattern by applying weight to the tool very slowly and allowing it to drill off. After making the first few feet in this way, approximately 500 lbs. of weight can be placed on the tool while observing pressure to detect stalling. Once pump pressure stabilizes and no stalling occurs, the weight can be increased to 700–800 lbs., again allowing the parameters to stabilize. Once this occurs, if drilling on a hard plug, the weight can be increased to 1,000–1,200 lbs. The pressure must be carefully observed during this time.

- The maximum penetration rate is normally determined by factors such as the weight on the bit at which the motor stalls, the material's hardness, the ability to clean the hole, and other parameters. Typical underreaming rates in cement in recent squeezes are approximately 20–40 ft./h, but this can vary significantly depending on the hardness. Cement that has been in place for several months can drill as slowly as 2–3 ft./h. Scale can usually be reamed at 30–60 ft./h, but harder sections may drill as slowly as 1–2 ft./h.

- The amount of weight on the bit is an important parameter. Typically, 500–1,000 lbs. is maintained when drilling cement or scale. If the tool stalls frequently, it may be necessary to apply as little as 100–200 lbs. for short intervals. In highly deviated holes, it can be difficult to determine the actual weight on the bottom from the weight indicator. Constant pump pressure, rather than the weight on the bit, must be maintained.

- Stalls shorten the life of the motor and reduce average penetration rates. A stall is indicated by a rapid increase in pump pressure without a corresponding increase in pump rate. When a stall occurs, pumping should be stopped and the pressure allowed to bleed off, ensuring that the motor has stopped. Before restarting the pump, the underreamer should be picked up at 10–30 ft. This reduces the potential for backing off at a connection in the BHA. Only after the pump pressure is stable should the underreamer be returned to the restriction depth.

- The relatively small fluid passages through the underreamer can get plugged easily if sand or other solids are introduced into the coiled tubing. To avoid plugging the passages, the power fluid—normally seawater or freshwater—should be filtered.

- Cuttings must be circulated out of the wellbore, which requires solid suspension and a fluid velocity in the annulus that is greater than the solids' settling velocity. Production from the well can assist in carrying out solids and keeping the cuttings from plugging the perforations.

- High-viscosity polymer sweeps are often used to assist in hole cleaning. Typically, a polymer sweep is pumped after 40–50 ft. of underreaming (or more often if hole conditions dictate). After underreaming is completed, a polymer pill is pumped out of the circulating sub to remove cuttings that may have fallen to the low side of the hole. Pump pressure will increase when pumping high-viscosity sweeps down the coiled tubing.

- Once target depth has been reached with the underreamer, two additional passes should be made from the tubing tailpipe to the cleanout depth to ensure that all restrictions have been removed. After this, the dual-circulation sub should be opened, a final polymer sweep pumped, and the flow rate increased to the maximum to circulate out the polymer sweep at the highest possible rate. Once the polymer has exited the coiled tubing, begin pulling out of the hole while continuing to pump at the maximum rate to ensure that suspended solids are lifted from the well.

COILED-TUBING-CONVEYED TUBING AND DRILL-PIPE CUTTING

29

Until the early 1990s, few pipe-cutting operations were performed using coiled tubing as a means of conveyance. A drilling or workover rig with a jointed-pipe workstring was the primary choice for conveying mechanical and hydraulic pipe-cutting tools when wireline methods were not suitable—particularly in highly deviated and horizontal wells. As its capabilities grow, coiled tubing is increasingly used with other tools to perform more challenging jobs, such as cutting single- and multiple-pipe strings.

HYDROMECHANICAL PIPE-CUTTING SYSTEM

When conveyed on coiled tubing to sever pipe, the hydromechanical pipe cutter is run below a workover motor. The workover motor provides the high torque at low speed required to perform the cut. This setup forms part of a hydromechanical pipe-cutting system (Figure 29-1).

The hydromechanical pipe cutter is composed of two major assemblies (Figure 29-2). The inner assembly connects to the string and consists of a top sub, mandrel, nozzle carrier, and knives. The outer assembly consists of a drive sleeve, upper body, lower body, and bottom plug. The two assemblies are interlocked, slide against each other, and are torque-locked by drive pins.

When weight or internal pressure is applied to the pipe-cutting tool, the pressure acts against an internal pressure cavity to push the outer assembly upward, at which point weight on the bit physically pushes it up (Figure 29-3). A spring allows it to return. The lower body of the tool incorporates a ramp that kicks out the cutting knives at a 45° angle when the tool is activated. The knives remain in the locked position so long as weight or internal pressure is applied to the tool. The cutter has three blades, making it self-stabilizing. This is particularly important in severely deviated wells. The tool is equipped with replaceable nozzles that allow flow and activation rates to be customized to specific well conditions. The cutter also incorporates a ball and probe to give the operator a positive pressure indication of a full knife extension.

The hydromechanical pipe cutter uses several cutting-blade configurations designed to cut smoothly through pipe of common dimensions and material compositions. Blades are also available for nonstandard material, such as metal containing more than 13% chrome. The cutting blades contain renewable cutting inserts, each of which is placed in a specific pattern to ensure that a new cutting insert will be exposed to the pipe wall if the previous one becomes worn. The blades are designed to suit the applied weight and torque available from the coiled tubing and workover motor. Cuttings typically are small, uniform, and easy to circulate out of the hole.

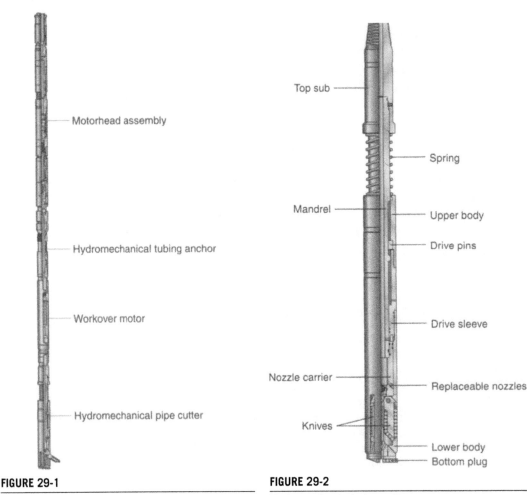

FIGURE 29-1

Hydromechanical pipe-cutting system.

(courtesy of Baker Oil Tools)

FIGURE 29-2

Hydromechanical pipe cutter.

(courtesy of Baker Oil Tools)

CENTRALIZATION AND STABILIZATION

Although the hydromechanical pipe cutter is self-stabilizing, residual bend of the coiled tubing and lateral forces generated in a deviated or horizontal wellbore can impede the success of a pipe-cutting operation.

A hydraulically actuated centralizer is typically placed between the dual-activated circulating sub and the down-hole workover motor. This provides lateral support to the upper part of the cutting assembly. A second, hydraulically actuated centralizer with a rotating inner mandrel can also be placed between the workover motor and the cutter. This helps to further stabilize and centralize the cutter during the pipe-cutting sequence. Using a lower-placed centralizer also ensures that all the cutting knives sustain the same amount of point loading, which can reduce motor stalls and improve the quality of the cut.

The hydraulically actuated centralizer is run into the hole with its bow springs retracted to allow easy passage through small diameters within the wellbore. Pumping through the centralizer causes the springs to extend and contact the pipe wall, which centers the cutting assembly in the wellbore. Back pressure to activate the centralizer can be achieved using an integral choke or using back pressure from the workover motor.

HYDROMECHANICAL TUBING ANCHOR

The hydromechanical tubing anchor (Figure 29-4) is used to eliminate movement of the coiled tubing while making the cut. The anchor uses a cone and collet to anchor the tool to the tubing. The collet is attached to a piston driven by hydraulic pressure. Mechanical downward force applied by setting down weight onto the coiled tubing holds the anchor in place while the cut is made.

COMPRESSION SWIVEL AND BULLNOSE

A compression swivel and bullnose sub are used when a restriction or bridge plug is located inside the pipe in the area where the cut is to be made. The compression swivel and bullnose are attached below the cutter and used to tag the bridge plug or restriction. A small amount of set-down weight is then applied, which places the assembly in compression. The cutter is activated, and the cut is made without the coiled tubing moving during the cut.

FIGURE 29-3

Hydromechanical pipe cutter in the open position.
(courtesy of Baker Oil Tools)

COMPUTER SOFTWARE

Software programs (Figure 29-5) are available to model the flow and operational characteristics of the hydromechanical pipe-cutting system before and during field operations. Doing this aids in determining the orifice size that will maximize the available hydraulic power to operate the cutting system. The software is also used to correlate the optimum surface-cutting speed for the pipe to be cut with the output of the workover motor.

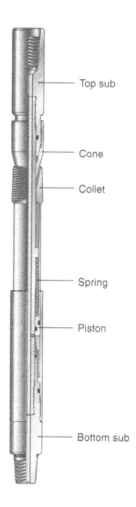

Top sub

Cone

Collet

Spring

Piston

Bottom sub

FIGURE 29-4

Hydromechanical tubing anchor.

(courtesy of Baker Oil Tools)

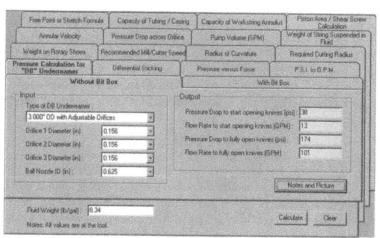

FIGURE 29-5

Hydromechanical pipe-cutting software program.

(courtesy of Baker Oil Tools)

THRU-TUBING IMPACT DRILLING

More than half of all coiled-tubing jobs are wellbore cleanouts. A significant number of these operations are performed using workover motors and milling assemblies. While the value of a milling assembly is well documented, down-hole conditions such as high temperatures (above 400 °F) and hostile fluids can significantly reduce the life of a workover motor and limit milling operations. Impact-drill systems (Figure 30-1) are an attractive alternative because they perform reliably in adverse conditions. The most common applications of the impact drill include scale milling, hard-cement milling, resin-sand removal, and gravel removal.

Impact drills are different from positive displacement motors (PDMs) in the following ways:

- They do not operate until the bit meets resistance.
- They do not store reverse torque.
- They can operate in high temperatures (above 600 °F).
- They will operate in most fluid media, including those with high volumes of nitrogen.
- They have short makeup lengths (3.6 ft. is average).
- They can be equipped to operate in hostile conditions (such as hydrogen sulfide or hydrochloric acid).
- They have low redress costs (as they are not constructed of elastomers).

Fluid pumped through the coiled tubing causes the impact drill to reciprocate and rotate at the same time. The frequency of the stroke depends on the amount of weight applied and the volume of fluid being pumped through the tool. The drill does not begin operating until the bit has met resistance, which permits circulation while running in and out of the well and prevents damage to the tool or tubing wall.

OPERATIONAL CONSIDERATIONS

- The amount of torque transmitted to the bit of an impact drill is a function of the amount of compressive load applied to the tool. A rule of thumb for this calculation is that approximately 15% of the compressive force is translated into torque.
- Operating pressure varies as a result of tool load. The more compressive force applied to the tool, the more pressure is required to cycle it. The impact drill will produce operating pressures ranging from 450 to 2,100 psi. This corresponds to an applied load range of 800–2,400 lbs.
- Optimum penetration rates are most often achieved at light bit loads and low circulation rates; 600–800 lbs. weight-on-tool (WOT) values and 25–50 gallons per minute are typical. It is tempting to load the tool with tubing weight and watch the unit shake. While this may be

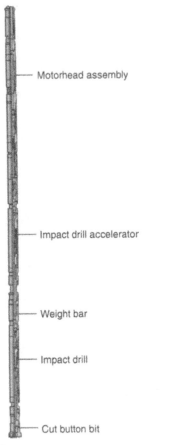

FIGURE 30-1

Impact-drilling bottomhole assembly.

(courtesy of Baker Oil Tools)

Motorhead assembly

Impact drill accelerator

Weight bar

Impact drill

Cut button bit

entertaining, it is usually not productive. If high circulation rates are necessary to clean out after drilling an interval, lift up the bottom by 1 ft. or so and circulate through the open tool.

- When running the impact drill using mud as the power fluid, the impact will be sluggish and penetration slow. Seat erosion will limit tool life. If possible, do not circulate the drill in and out of the hole.
- Never run a device on the bottom of the tool that would restrict fluid exit. The exhaust-port area should be at least double the flow area through the tool.

BIT DESIGN

It is always advisable to run bits with specially designed impact-grade inserts. Standard tungsten carbide inserts cannot withstand the shock loads generated at the bit by the impact drill. The most common types of bits used in impact drill applications are the following:

- A cut-button bit (Figure 30-2A) is the optimum impact bit to use on cement, rock, or hard scales. It is also effective in removing sand and other types of fill.
- A six-point carbide bit (Figure 30-2B) works well on resin sands, hard-packed sand, and fill.
- A ceramic disk breaker (Figure 30-2C) is a pointed, blind box used to fracture ceramic or glass isolation disks in completions.

Drilling with an impact force yields the benefit of first cracking (or at least introducing stress lines) in scale or cement, which is followed by high-torque bit rotation and high-pressure pulses from the fluid medium. Returns often contain chips with a convex surface molded by the tubing wall, indicating that impact forces have thoroughly cleaned the wellbore.

PENETRATION RATES

As with milling and underreaming jobs, impact-drilling penetration rates can vary greatly depending on the amount and hardness of the scale or cement. In extreme cases, rates can be as slow as 3 ft./h (usually through the upsets). An acceptable penetration rate in badly scaled wells is 80–100 ft./h.

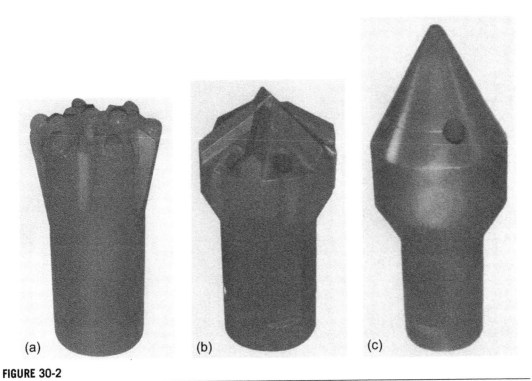

(a)   (b)   (c)

FIGURE 30-2

(A) Cut-button bit; (B) six-point carbide bit (C) ceramic disk breaker.

(All images courtesy of Baker Oil Tools.)

It may be more efficient to open the wellbore with an undersized bit and then make a second (or cleanup) run with a cleanup bit of the desired diameter. Leaving too thin a layer on the tubing walls can slow penetration rates on the second run because the bit is more likely to jam. Leaving a ⅛-in. or thicker layer for the cleanup run helps to prevent this problem.

PRECAUTIONARY MEASURES AND INSPECTION PROCEDURES

Handling all down-hole tools in the correct manner is a major measure to prevent problems. Do not overlook the fact that small actions can have the desired effects. Examples of effective procedures include the following:

- Use the recommended slips for the tools being used. Longer slips with larger contact areas may be needed for heavier loads depending on the well depth.
- Try to always use slip-type elevators and double elevators for tools built with higher-grade steel.
- Always use a drill collar (safety) clamp on slick outside diameters (ODs) or tools without an upset tool joint.
- Watch for equipment wear. Always make the connections using the correct techniques and torque. If a freshly disconnected joint has mud in it, clean it thoroughly and check for damages or washouts.
- During fishing operations, always replace the fishing assembly, especially the jars, bumper subs, and accelerators.

Most of these recommendations are common practices, but they may be overlooked in the distraction that can happen during fishing operations. Don't be complacent about your role on location—this can lead to an additional fishing operation, putting a fish on top of a fish.

Tool inspection is a common practice for drill strings, and this typically done after a certain number of hours of drilling. However, when it comes to fishing tools, this practice should be followed after each fishing job when the tools are sent back to the warehouse for service. At a minimum, always use a qualified company to complete a magnetic particle inspection on key components or the areas as outlined, based on the practices for service and inspection recommended by the tool manufacturer.

SERVICE PROVIDERS

The following list contains the major global service companies that provide down-hole fishing equipment that is sold or can be obtained on a rental basis. There are also other smaller service companies not listed here. All these firms sell a wide variety of down-hole fishing equipment and employ quality fishing tool supervisors with years of experience in the industry.

These providers are constantly updating their technology and products, so check with them for the latest equipment for your downhole fishing needs. If you are interested in finding local providers in your area, check the Internet.

Baker Hughes, Inc.
17015 Aldine Westfield Rd.
Houston, TX 77073-5101
(713) 625-4200
www.bakerhughes.com
Fishing, casing exits, milling services, thru-tubing services, wellbore cleaning, external casing patches

Weatherford International, LLC
2000 St. James Place
Houston, TX, 77056
(713) 836-4000
www.weatherford.com
Fishing, casing exits, milling services, thru-tubing services, wellbore cleaning, internal and external casing patches

Schlumberger Oilfield Services
1325 S. Dairy Ashford Dr.
Houston, TX 77077
(281) 285-1300
www.slb.com
Fishing, casing exits, milling services, thru-tubing services, external casing patches

National Oilwell Varco
7909 Parkwood Circle Dr.
Houston, TX 77036
(713) 634-3311
www.nov.com
Fishing, jars, agitators, milling equipment, thru-tubing fishing equipment, external casing patches

Login Oil Tools
Remington Square Office Building
10603 W. Sam Houston Parkway N.
Suite 200
Houston, TX 77064-4362
(832) 386-2500
www.loganoiltools.com
Fishing, jars, milling equipment, thru-tubing fishing equipment, external casing patches

GOTCO International, Inc.
11410 Spring Cypress Rd.
Tomball, TX 77377
(281) 376-3784
www.gotco-usa.com
Fishing, jars, milling equipment, external casing patches

EV Downhole Video
15720 Park Row
Suite 500
Houston, TX 77084
(281) 492-1300
www.evcam.com
Downhole Video cameras

DOWN-HOLE VIDEO CAMERAS

33

In today's world of downhole fishing, we have the ability to utilize technology to give us the highest possibility of success to capture the unknown fish or see problems downhole that an impression block or tale-tale signs from surface indications may not be easily understood by the fishing tool operator. A couple of examples (Figure 33.1) is a dual string where one of the tubing strings has parted and dropped down hole which can be clearly seen on the video caption. The second caption (Figure 33.2) is where the tubing was attempted to be washed over and the mill had went beside the tubing as seen by the mill marks (grooves) on the side of the pipe.

The introduction of down-hole video cameras (DHVs) in both memory and live feed has given the fishing tool operator a whole new look into the operation of fishing. A wide range of cameras can be deployed on e-line, slick line, coil tubing, and drill pipe; these devices use state-of-the-art video, lighting, and communications technology. Cameras are rated to operate at temperatures up to 257°F and 15,000 psi with 1 11/16" OD in diameter. Some cameras can even work at temperatures of up to 350°F and pressures of up to 22,500 psi, but very few are available globally today.

The memory camera records for up to 5 h in color and in high-definition, and at a speed of 30 frames per second. It is a time-delayed camera with up to 10 programmable time intervals. It is powered by a lithium-ion battery, which can last over up to 24 h. The memory camera is primarily run on wireline or coil tubing. It can be run with other logging tools so long as it is at the bottom of the string, which allows quantitative and qualitative data to be captured.

The real-time, E-line camera also records in color and operates at up to 25 frames per second with down- and side-view lenses, and a live feed can be linked to an engineer via the Internet. The E-line camera is completely digital, and it can change its lighting intensity, resolution, and rotation, and it can switch to side view, all from the surface. The E-line is a modular system that allows for adjusting the length and deciding what tools are needed based on the application. Operators can also run a live, 24-arm caliper that works in tandem with the E-line camera, which allows it to broadcast a live visual of any anomalies in the caliper log.

Wellbore cleanup methods are a key to the successful run of DHVs. There are numerous well cleanup methods that can ensure the highest possible success rate. Please refer to SPE paper 35680 for wellbore cleanup depending on your wellbore and application, or contact a local DHV provider.

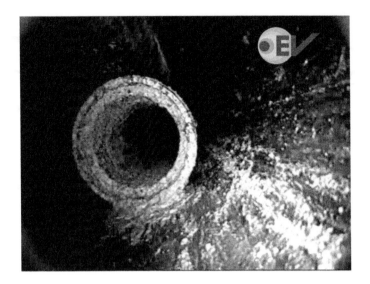

FIGURE 33-1

Top of parted pipe.

(courtesy of EV Downhole)

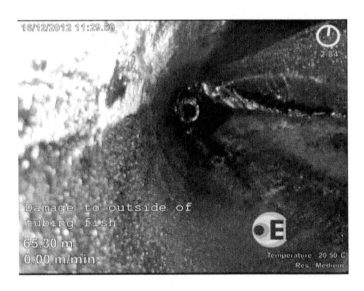

FIGURE 33-2

Damage to outside of Tubing.

(courtesy of EV Downhole)

Glossary

Accelerator (intensifier) A down-hole tool used in conjunction with a jar to store energy that is released suddenly when the jar operates.

Annular velocity The linear velocity of a fluid passing through an annular space. The term *critical annular velocity* is often used to describe the flow rate or velocity at which entrained solids will be efficiently transported by the annular fluid. If the fluid velocity falls below the critical rate, there will be a risk of particles settling, forming beds or bridges that may obstruct the wellbore.

Back off To unscrew one threaded piece (as a section of pipe) from another.

Back pressure The pressure within a system caused by fluid friction or an induced resistance to flow through the system. Hydraulic thru-tubing fishing tools are typically operated using back pressure created by flow through a nozzle.

Bailer A long, cylindrical container, fitted with a valve at its lower end, used to remove water, sand, mud, oil, and debris from a well.

Bent sub A short cylindrical device installed in a drill stem between the bottommost drill collar and a down-hole mud motor. The purpose of the bent sub is to deflect the workover motor off vertical to drill a directional hole.

Blowout preventer (BOP) A large valve at the top of a well that may be closed if the drilling crew loses control of formation fluids. By closing this valve (usually operated remotely via hydraulic actuators), the drilling crew usually regains control of the reservoir, and procedures can then be initiated to increase the mud density until it is possible to open the BOP and retain pressure control of the formation.

Bottom-hole assembly (BHA) The tools run below the work-string of pipe or coiled tubing.

Bridge An obstruction in the bore hole, caused by a buildup of material such as scale, wellbore fill, or cuttings, that can restrict wellbore access or, in severe cases, eventually close the wellbore. In an open hole, a bridge usually caused by the caving in of the wall of the bore hole or by the intrusion of a large boulder.

Buoyancy The upward force acting on an object placed in a fluid. The buoyancy force is equal to the weight of fluid displaced by the object. Buoyancy can have significant effects over a wide range of completion and work-over activities, especially in cases in which the wellbore and tubing string contain liquid and gas. Any change in the relative volumes or fluid levels will change the buoyancy forces.

Bushing A pipe fitting that allows two pieces of pipe of different sizes to be connected.

Cased hole The portion of the wellbore that has had metal casing placed and cemented to protect the open hole from fluids, pressures, wellbore stability problems, or a combination of these.

Catcher A device fitted into a junk basket and acting as a trapdoor to retain the junk.

Centralizer A tool used to keep a tool string in the center of the tubing, casing, or wellbore. Tool centralization may be required to prevent the tool from hanging up on obstructions on the wellbore wall, to place fluid efficiently, and to avoid excessive standoff.

Coiled tubing (1) A long, continuous length of pipe wound on a spool. The pipe is straightened prior to pushing into a wellbore and is recoiled to spool the pipe back onto the transport and storage spool. (2) A generic term relating to the use of a coiled tubing string and associated equipment. As a well-intervention method, coiled tubing techniques offer several key benefits over alternative well-intervention technologies. The ability to work safely under live well conditions, with a continuous string, enables fluids to be pumped at any time, regardless of the position or direction of travel. This is a significant advantage in many applications.

Coiled tubing fatigue-cycle life A term referring to the fact that the useful life of a coiled tubing work-string is limited due to fatigue damage caused by the repeated bending and straightening of the coiled tubing at the gooseneck and reel. The resulting failure mechanism is referred to as *low cycle fatigue*. Tubing damage increases dramatically if internal pressure is applied while the coiled tubing is bent. The coiled tubing fatigue-cycle life can often be a limiting factor in the fishing job.

Coiled tubing unit The package of equipment required to run a coiled tubing operation. Four basic components are required: (1) the coiled tubing reel to store and transport the coiled tubing string, (2) the injector head to provide the tractive effort to run and retrieve the coiled tubing string, (3) the control cabin from which the equipment operator controls and monitors the operation, and (4) the power pack that generates the necessary hydraulic and pneumatic power required by the other components. Pressure-control equipment is incorporated into the equipment to provide the necessary control of well pressure fluid during normal operating conditions and contingency situations requiring emergency control.

Collar A coupling device used to join two lengths of pipe. A combination collar has different threads on each end.

Collar locator A logging device for depth-correlation purposes, operated mechanically or magnetically to produce a log showing the location of each casing, tubing collar, and coupling in a well. It provides an accurate way to measure the depth of a well.

Completion fluid A special drilling mud used when a well is being completed. It is selected not only for its ability to control formation pressure, but also for its properties that minimize formation damage.

Crooked hole A wellbore that has deviated from the vertical. Crooked holes usually occur where there is a section of alternating hard and soft strata steeply inclined from the horizontal.

Cuttings Small pieces of rock, cement, or steel that break away due to the action of the bit, milling, cutting, or underreaming tools.

Deviated wellbore A wellbore that is not vertical. This term usually indicates a wellbore that has been drilled away from vertical intentionally.

Dressing A term used to describe the fitting together of all parts of a tool or the surfacing, or a tool with particular materials, such as "dressing" a mill with carbide.

Dutchman A piece of tubular pipe that has been broken or twisted off a female connection. It may also continue past the connection.

Fish Any object in a well that obstructs drilling or operation; usually consists of pipe or junk.

Fishing The use of tools, equipment, and techniques to remove junk and debris from a wellbore. The key elements of a fishing operation include an understanding of the dimensions and nature of the fish to be removed, the wellbore conditions, the tools and techniques employed, and the process by which the recovered fish will be handled at the surface.

Fishing neck The surface on which a fishing tool engages (internally or externally) when retrieving tubing, tools, or equipment stuck or lost in a wellbore. Tools and equipment that are temporarily installed in a wellbore are generally equipped with a specific fishing-neck profile to enable the running and retrieval tools to reliably engage and release the objects in question.

Flush-joint pipe Pipe in which the outside diameter (OD) of the joint is the same as the OD of the tube. It may also be flush-jointed internally.

Free-point The depth at which pipe is stuck; more specifically, the depth immediately above the point at which the pipe is stuck.

Friction reducer An additive, generally in liquid form, used to reduce the friction forces of fluid circulation.

Gooseneck An assembly mounted on a coiled tubing injector head that guides the tubing string as it passes through an arc from the reel into a vertical alignment with the injector-head chains and wellbore. The radius of the guide arch is generally designed to be as large as practicable because the plastic deformation created in the coiled tubing string induces material fatigue in the tube.

Grapple The part of a catching tool (such as an overshot or spear) that engages the fish.

Horizontal drilling Drilling in which the departure of the wellbore from vertical exceeds about 80 degrees. Because a horizontal well typically penetrates a greater length of the reservoir, it can offer significant production improvement over a vertical well.

Hydrostatic head The pressure exerted by a body of liquid at rest. The hydrostatic head of fresh water is 0.433 psi per foot of height. For other liquids, it may be determined by comparing their specific gravities with the specific gravity of water.

Hydrostatic pressure The pressure at any point in a column of fluid caused by the weight of fluid above that point.

Impression block A tool made of a soft material such as lead or coal tar and used to secure an imprint of a fish.

Injector head One of the principal equipment components of a coiled tubing unit. The injector head incorporates special profiled chain assemblies to grip the coiled tubing string, and uses a hydraulic drive system that provides the traction for running and retrieving the string from the wellbore. The base of the injector head is secured to the wellhead pressure-control equipment by the stripper assembly mounting system. The gooseneck mounted on top of the injector head feeds the tubing string from the reel around a controlled radius into the injector head.

Jar A down-hole tool used to deliver an impact force either up or down on the tool string, usually to operate down-hole tools or to dislodge a stuck tool string. Jars of different designs and operating principles are commonly included on slick-line, coiled tubing, and workover tool strings. Simple slick-line jars incorporate an assembly that allows some free travel within the tool to gain momentum for the impact that occurs at the end of the stroke. Larger, more complex jars for coiled tubing or workover strings incorporate a trip or firing mechanism that prevents the jar from operating until the desired tension is applied to the string, optimizing the impact delivered. Jars are designed to be reset by simple string manipulation and are capable of repeated operation or firing before being recovered from the well.

Junk Metal debris lost or left in a wellbore. Junk may be a bit, cones from a bit, hand tools, or any small obstruction.

Junk basket A cylindrical tool designed to retrieve junk or foreign objects loose in a wellbore.

Junk sub (boot basket) A tool run just above the bit or mill in the drill stem to catch small, nondrillable objects circulating in the annulus.

Key seat A channel or groove cut in the side of a hole parallel to the axis. Key seats result from the dragging of pipe on a sharp bend in the hole.

Kick An entry of water, gas, oil, or other formation fluid into the wellbore. It occurs because the pressure exerted by the column of drilling fluid is not great enough to overcome the pressure exerted by the fluids in the formation drilled. If prompt action is not taken to control the kick or kill the well, a blowout will occur.

Kill weight fluid A mud whose density is high enough to produce a hydrostatic pressure at the point of influx in a wellbore and shut off flow to the well.

Liner Any string of casing whose top is located below the surface. A liner may serve as the oil string, extending from the producing interval up to the next string.

Macaroni string A string of tubing of a very small diameter.

Magnet A permanent magnet or electromagnet fitted into a tool body so that it may be run to retrieve relatively small junk made of ferrous metal.

Make up To connect tools or tubulars by assembling the threaded connections incorporated at either end of every tool and tubular. To do this, the threaded tool joints must be correctly identified and then torqued to the correct value to ensure a secure tool string and connection pressure integrity.

Mandrel A cylindrical bar, spindle, or shaft around which other parts are arranged or attached, or that fits inside a cylinder or tube.

Measure in To obtain an accurate measurement of the depth reached in a well by measuring the drill pipe or tubing as it is run into the well.

Measure out To measure drill pipe or tubing as it is pulled from the hole, usually to determine the depth of the well or the depth to which the pipe or tubing is run.

Mill A down-hole tool with rough, sharp, extremely hard cutting surfaces for removing metal by cutting. Mills are run on drill pipe or tubing to cut up debris in the hole and to remove stuck portions of the drill stem or sections of casing for sidetracking.

Milling The use of a mill or similar down-hole tool to cut and remove material from equipment or tools located in the wellbore. Successful milling operations require the appropriate selection of milling tools, fluids, and techniques. In these operations, mills or similar cutting tools must be compatible with the fish materials and wellbore conditions. The circulated fluids should be capable of removing the milled material from the wellbore.

Multiple completion An arrangement for producing a well in which one wellbore penetrates two or more petroleum-bearing formations that lie one over the other. The tubing strings are suspended side by side in the production casing string, each a different length and packed off to prevent the commingling of different reservoir fluids. Each reservoir is then produced through its own tubing string.

Necking The tendency of a metal bar or pipe to taper to a reduced diameter at some point when subjected to excessive longitudinal stress.

Nipple A completion component fabricated as a short section of heavy wall tubular with a machined internal surface that provides a sealing area and a locking profile. Landing nipples are included in most completions at predetermined intervals to enable the installation of flow-control devices, such as plugs and chokes. Three basic types of landing nipple are commonly used: no-go nipples, selective-landing nipples, and ported or safety-valve nipples.

Nitrogen-based fluid A multiphase fluid incorporating a liquid base and gaseous nitrogen. Nitrified fluids are often used in stimulation treatments to enhance the performance of the treatment fluid and improve the cleanup process following the treatment.

Overpull Pull on pipe over and above its weight, in either air or fluid.

Perforate To create holes in a casing or liner to achieve efficient communication between the reservoir and the wellbore. A perforating gun assembly with the appropriate configuration of shaped explosive charges and the means to verify or correlate the correct perforating depth can be deployed on wireline, tubing, or coiled tubing.

Pill A relatively small volume of specially prepared fluid placed or circulated in the wellbore. Fluid pills are commonly prepared for a variety of special functions, such as a sweep pill prepared at high viscosity to circulate around the wellbore and pick up debris or wellbore fill. To counteract lost-circulation problems, a lost-circulation pill prepared with flaked or fibrous material is designed to plug the perforations or formation interval, losing the fluid.

Polycrystalline diamond compact (PDC) A cutting structure made of synthetic diamond.

Positive displacement motor (PDM) A motor that uses hydraulic horsepower of the drilling fluid, based on the Moineau principle, to drive the drill bit. Workover motors are used extensively in coiled tubing applications requiring rotation, such as milling, underreaming, and cutting.

Pressure deployment system An assembly of pressure-control equipment that enables the running and retrieval of long tool strings on a coiled tubing string in a live wellbore. The deployment system is configured to provide two barriers against well pressure as the tool string is assembled and run into the wellbore. Once fully assembled, the coiled tubing equipment is connected and the tool string is run into the wellbore. The process is reversed for tool retrieval.

Residual bend The natural form that a section of coiled tubing string will take if spooled from the reel and allowed to rest without any tension applied. The residual bend results from the plastic deformation imparted as the string is spooled around the radius of the reel and guide arch.

Reverse-circulate To pump down the annulus and back up the work-string (drill pipe or tubing). This action is often used for workover in cased holes.

Rig down To take apart equipment for storage and portability. Equipment typically must be disconnected from power sources, decoupled from pressurized systems, disassembled, and moved off the rig floor, or even off site.

Rig up To make equipment ready for use. Equipment must typically be moved onto the rig floor, assembled, and connected to power sources or pressurized piping systems.

Riser (lubricator) A term initially applied to the assembly of pressure-control equipment used on slick-line operations to house the tool string in preparation for running into the well or for retrieving the tool string upon completion of the operation. The lubricator is assembled from sections of heavy-wall tube generally constructed with integral seals and connections. Lubricator sections are routinely used in the assembly of pressure-control equipment for other well-intervention operations, such as coiled tubing.

Rotary shoe The cutting shoe fitted to the lower end of washover pipe and "dressed" with hard-surfaced teeth or tungsten carbide.

Safety joint A connection with coarse threads or other special features that will cause it to unscrew before other connections in the string.

Sand line A wire rope used on well-servicing rigs to operate a swab or bailer. It is usually $\frac{9}{16}$ in. in diameter and several thousand feet long.

Sinker bar A heavy weight placed on or near a lightweight wireline tool. It adds weight so that the tool can be lowered into the well properly.

Sliding sleeve A completion device that can be operated to provide a flow path between the production conduit and the annulus. Sliding sleeves incorporate a system of ports that can be opened or closed by a sliding component that is generally controlled and operated by slick-line or coiled tubing tool string.

Spear A tool that goes inside a tubular fish and catches it with a slip.

Squeeze cementing The forcing of cement slurry via pressure on specified points in a well to cause seals at the squeeze points. It is a secondary cementing method used to isolate a producing formation, seal off water, repair casing leaks, and other similar actions.

Stinger Any cylindrical or tubular projection, relatively small in diameter, that extends below a down-hole tool and helps to guide it to a designated spot (such as the center of a portion of stuck pipe).

String The entire length of casing, tubing, or drill pipe run into a hole.

String shot An explosive line which, when detonated, delivers a concussion to pipe, causing it to unscrew or "back off." Also called *Prima-Cord*.

Substitute (Sub) A short section of pipe, tube, or drill collar with threads on both ends, used to connect two items with different threads; an adapter.

Surface pipe The first string of casing set in a well after the conductor pipe, varying in length from a few hundred feet to several thousand feet. Some states require a minimum length to protect freshwater sands.

Surfactant A substance that affects the properties of the surface of a liquid or solid by concentrating on the surface layer. Reduces surface tension, thereby causing fluid to penetrate and increasing "wettability."

Swage (or swage mandrel) A tool used to straighten damaged or collapsed pipe in a well.

Tail pipe The tubulars and completion components run below a production packer. The tail pipe may be included in a completion design for several reasons. It can provide a facility for plugs and other temporary flow-control devices, improve down-hole hydraulic characteristics, and provide a suspension point for down-hole gauges and monitoring equipment.

Twist off To part or split drill pipe or drill collars, primarily because of metal fatigue.

Underbalanced The amount of pressure (or force per unit area) exerted on a formation exposed in a wellbore below the internal fluid pressure of that formation. If sufficient porosity and permeability exist, formation fluids enter the wellbore. The drilling rate typically increases as an underbalanced condition is approached.

Underream To enlarge or open up the wellbore below the casing or a restriction.

Viscosity A property of fluids and slurries that indicates their resistance to flow, defined as the ratio of shear stress to shear rate.

Washover pipe (washpipe) Pipe of an appropriate size to go over a fish in an open hole or casing and wash out or drill out the obstruction so that the fish may be freed.

Well schematic A schematic diagram that identifies the main completion components installed in a wellbore. The information included in the wellbore diagram relates to the principal dimensions of the components and the depth at which the components are located. A current wellbore diagram should be available for any well intervention operation to enable engineers and equipment operators to select the most appropriate equipment and prepare operating procedures that are compatible with any down-hole restrictions.

Wire line A general term used to describe well-intervention operations conducted using single-strand or multistrand wire or cable for intervention in oil or gas wells. Although applied inconsistently, the term is used commonly in association with electric logging and cables incorporating electrical conductors. Similarly, the term *slick-line* is commonly used to differentiate operations performed with single-strand wire or braided lines.

Workover The process of performing major maintenance or remedial treatments on an oil or gas well. In many cases, workover implies the removal and replacement of the production tubing string after the well has been killed and a workover rig has been placed on location. Thru-tubing workover operations, using coiled tubing, snubbing, or slick-line equipment, are routinely conducted to complete treatments or well service activities that avoid a full workover where the tubing is removed. This operation saves considerable time and expense.

Yield point The yield stress extrapolated to a shear rate of zero.

BIBLIOGRAPHY

I would like to thank Baker Hughes Incorporated for allowing me to use illustrations from their catalogs, technical papers, and brochures. More information on these materials, and about the company in general, can be found at www.bakerhughes.com.

Thanks to National Oilwell for the use of illustrations from their catalogs, which can be found at www.nationaloilwell.com.

We also give thanks to Logan Oil Tools for supplying illustrations used in this book. You can find additional information at www.loganoiltools.com.

EV Downhole Video was also very helpful, supplying technical information and captions for their downhole video cameras. Check the company's website at www.evcam.com.

BOOKS, PAPERS, AND ARTICLES OF INTEREST

Adkins, C. S. (1993). "Economics of Fishing". SPE Technology #20320. JPT (May).*

Coronado, M. P., Baker Oil Tools. "Coiled Tubing Conveyed Fishing Systems". Presented at the World Oil 1993 Coiled Tubing Operations and Slimhole Drilling Practices Conference.

Fanguy, D. J. "Coiled Tubing Conveyed Hydromechanical Pipe Cutting: A Safe, Effective Alternative to Chemical and Explosive Severing Methods". Presented at the SPE ICoTA 2001 Coiled Tubing Roundtable and Exhibition.*

Forster, I., Grant, R., 2012. Axial Excitation and Drill String Resonance as a Means of Aiding Tubular Retrieval: Small-Scale Rig Testing and Full-Scale Field Testing", SPE 151096, San Diego.

Going, W. S., and D. B. Haughton, Baker Oil Tools. "Using Multi-Function Fishing Tool Strings to Improve Efficiency and Economics of Deepwater Plug and Abandonment Operations". SPE/IADC Paper #67747.*

Harrison, G., July 1980. Fishing Decisions Under Uncertainty. J. Petrol. Tech.

Hinojosa, R., Texaco North America Production; J. Ryan and R. Wyman, Baker Oil Tools; and B. Wiley, Texaco North America Production. "Whipstock Performance Review in Gulf Coast Region Yields Operational." IADC/SPE Paper 39402. March 1998.*

Kemp, G., 1990. Oilwell Fishing Operations: Tools and Techniques, second ed. Gulf Publishing Company, Houston.

Lambertus, C. F., G. M. Joppe, and M. McGurk (2006). "Using High-Frequency Downhole Vibration Technology To Enhance Through-Tubing Fishing and Workover Operations", SPE Paper #99414.

Mohanna, A., C. Hanley, A. Mousa, and A. Al-Amri (2013). "Downhole Vibration Analysis: Fishing Agitation Tool Efficiency in Stuck Pipe Recovery", SPE-163516-MS, the Netherlands.

Nazzel, G., H. Rehbock, and T. Miller, Baker Oil Tools (1996). "Development, Testing and Field History of a True One Trip Casing Exit System". SPE Paper #35662. May.*

Short, J.A., 1995. Prevention Fishing and Casing Repair.

Stoesz, C. W., and J. P. DeGeare, Baker Oil Tools. "Low-Frequency Downhole Vibration Technology Applied to Fishing Operations". SPE Paper #63129.*

Tybero, P., Amoco Norway; and T. F. Bailey and S. Billeaud, Smith International (1996). "Technical Innovations Cut Sidetracking Time/Cost". SPE Paper #36671. October.*

Voghell, M., A. Mohanna, C. Hanley, C. Al-Khriseh, A. Mousa, A. Al-Amri (2013). "Downhole Vibration Analysis: Fishing Agitation Tool Efficiency in Stuck Pipe Recovery", SPE 163516-MS, the Netherlands.

Walker, G. (1984). "Fishing". SPE Paper #13360. October.*

Welch, James L., Whitlow, Richard R., 1988. Underreaming. In: World Oil's Coiled Tubing Handbook. third ed. Gulf Publishing Company, Houston.

Whittaker, J. L., and G. D. Linville, "Well Preparation—Essential to Successful Video Logging", SPE 35680.

Index

Note: Page numbers followed by *f* indicate figures, and *t* indicate tables.

Printed and bound by CPI Group (UK) Ltd, Croydon, CR0 4YY

03/10/2024

01040326-0004